PHYSICAL METHODS IN
HETEROCYCLIC CHEMISTRY

VOLUME VI

PHYSICAL METHODS IN HETEROCYCLIC CHEMISTRY

A Comprehensive Treatise in Six Volumes

Volume I—Nonspectroscopic Methods; Covering nonspectroscopic methods to 1962

Volume II—Spectroscopic Methods; Covering spectroscopic methods to 1962

Volumes III and IV—Covering methods other than X-ray structure analysis from 1963 to 1970

Volume V—Bond Lengths and Angles from X-Ray Crystallography; Covering X-ray structure analysis to 1970

Volume VI—Covering five new physical methods (also includes Cumulative Subject and Author Indexes for Volumes I–VI).

Physical Methods in Heterocyclic Chemistry

Edited by

A. R. KATRITZKY

School of Chemical Sciences
University of East Anglia
Norwich, England

VOLUME VI

1974

ACADEMIC PRESS · New York and London

A Subsidiary of Harcourt Brace Jovanovich, Publishers

ACADEMIC PRESS, INC.
111 Fifth Avenue, New York, New York 10003

United Kingdom Edition published by
ACADEMIC PRESS, INC. (LONDON) LTD.
24/28 Oval Road, London NW1

Library of Congress Cataloging in Publication Data
Main entry under title:

Physical methods in heterocyclic chemistry.

    Includes bibliographies.
    1.   Heterocyclic compounds.   I.   Katritzky, Alan R.,
ed.   [DNLM: 1.   Heterocyclic compounds.   QD400 K19p]
QD400.P59   547'.59   63-12034
ISBN 0–12–401106–3 (v. 6)

D
547·59
KAT

Contents

· 1 ·

UV Photoelectron Spectra of Heterocyclic Compounds

E. HEILBRONNER, J. P. MAIER, AND E. HASELBACH

· 2 ·

Microwave Spectroscopy of Heterocyclic Molecules

J. SHERIDAN

· 3 ·

ESR Spectroscopy of Heterocyclic Radicals

B. C. GILBERT AND M. TRENWITH

CONTENTS

· 4 ·

Fluorescence and Phosphorescence of Heterocyclic Molecules

STEPHEN G. SCHULMAN

· 5 ·

Some Applications of Thermochemistry to Heterocyclic Chemistry

KALEVI PIHLAJA AND ESKO TASKINEN

Contributors to Volume VI

Numbers in parentheses indicate the pages on which the authors' contributions begin.

B. C. Gilbert (95), *Department of Chemistry, University of York, Heslington, York, England*

E. Haselbach (1), *Physikalisch-Chemisches Institut der Universität Basel, Basel, Switzerland*

E. Heilbronner (1), *Physikalisch-Chemisches Institut der Universität Basel, Basel, Switzerland*

J. P. Maier (1), *Physikalisch-Chemisches Institut der Universität Basel, Basel, Switzerland*

Kalevi Pihlaja (199), *Department of Chemistry, University of Turku, Turku, Finland*

Stephen G. Schulman (147), *College of Pharmacy, University of Florida, Gainesville, Florida*

J. Sheridan (53), *University College of North Wales, Bangor, U.K.*

Esko Taskinen (199), *Department of Chemistry, University of Turku, Turku, Finland*

M. Trenwith (95), *Department of Chemistry, University of York, Heslington, York, England*

Preface

This volume of "Physical Methods in Heterocyclic Chemistry" completes the treatise. Coverage is now extended to some of the more recently introduced physical methods which have rapidly become important to heterocyclic chemists.

Volumes I and II, published in 1963, covered nonspectroscopic and spectroscopic methods, respectively, up to 1962. Volumes III and IV, published in 1971, covered an additional eight years of work on these same methods. Originally Volume V, comprised of a review of heterocyclic X-ray structure compiled by Dr. Wheatley, was to have completed this work. Two factors caused an alteration in this plan: the amount of material on X-ray structure proved to be greater than expected and a group of more recently introduced physical methods rapidly became of such increasing importance to heterocyclic chemists that it was necessary to include them to give the treatise the comprehensive coverage desired. Thus this, the final volume, includes accounts of five additional physical methods not previously covered plus cumulative Subject and Author Indexes which include information for all the volumes.

The topics comprising this volume are photoelectron spectroscopy (E. Heilbronner, J. P. Maier, and E. Haselbach), microwave spectroscopy (J. Sheridan), electron spin resonance (B. C. Gilbert and M. Trenwith), fluorescence and phosphorescence (Stephen G. Schulman), and thermochemical applications to heterocyclic chemistry (Kalevi Pihlaja and Esko Taskinen).

Editing this work has been a challenging but most enjoyable task. I would like to thank the many persons who by their contributions have helped to bring this work to fruition.

A. R. KATRITZKY

Preface to Volumes I and II

Physical methods are perhaps the most important of all the influences which have contributed to the fundamental changes of the last 50 years in the theory and practice of organic chemistry. Effective chemical research can now hardly be carried out without the aid of a variety of physical measurements.

In the advance of physical techniques into organic chemistry, two main streams may be identified: physical chemists have commenced with the study of the simplest molecules and, using methods as rigorous as practicable, have proceeded stage by stage to structures of increasing complexity. Organic chemists have, by contrast, frequently made correlations of the (usually complex) structures with which they work: such correlations being, at least at first, purely empirical. Both streams are of vital importance to the over-all development—they complement each other, and chemists of each type need to be aware of the work in both streams.

The systematic application of physical methods to heterocyclic chemistry has been slower than that to the other two traditional divisions of organic chemistry. This is probably because the molecular complexity of the heterocyclic field has hindered the advance into it by the physical chemist. A result is that most reviews and expositions of a physical method, or of a group of physical methods, deal but cursorily with its application to compounds of the heterocyclic class. The present two volumes seek to fill this gap—each chapter gives but a brief outline of the general theoretical and experimental aspects of the subject, and then gets down to surveying the literature in which the method has been applied to heterocyclic problems. This literature is often voluminous and is nearly always scattered. It is hoped that the present collection of reviews will save individual research workers much time and effort in literature searching.

As Editor, I have been fortunate in being able to enlist an international team of authors who are among the leaders in their respective fields, and my thanks go to each of them for their cooperation. We have tried to cover the literature to the beginning of 1962.

A. R. KATRITZKY

Preface to Volumes III, IV, and V

Since the publication in 1963 of the first two volumes of this treatise, the application of physical methods to organic chemistry, and in particular to heterocyclic chemistry, has proceeded apace. The importance of physical methods to structure determination and to the understanding of inter- and intramolecular interactions has increased no less than the flood of new work. Heterocyclic chemists are thus faced with the necessity of having more to comprehend for the efficient execution of their work.

The favorable reception given to volumes I and II encouraged us to update the work. All the chapters comprising the first two volumes have (with two exceptions) been amended. In addition, six new chapters are included covering methods which have achieved importance since 1963.

Volume III includes the chapters on ionization constants (A. Albert) and on ultraviolet spectra (W. L. F. Armarego), topics included in Volumes I and II, respectively. Volume III also covers the following new topics: gas electron diffraction (P. Andersen and O. Hassel), Raman spectroscopy (G. J. Thomas, Jr.), mass spectrometry (G. Spiteller), gas–liquid chromotography (Ya. L. Gol'dfarb *et al.*), and optical rotatory dispersion, circular dichroism, and magnetic dichroism (R. B. Homer).

Volume IV includes chapters on electric dipole moments (J. Kraft and S. Walker) and heteroaromatic reactivity (J. H. Ridd), which originally appeared in Volume I, and chapters on nuclear quadrupole resonance (E. A. C. Lucken), nuclear magnetic resonance (R. F. M. White and H. Williams), and infrared spectra (A. R. Katritzky and P. J. Taylor), which originally formed part of Volume II. Volume IV also includes one new topic: dielectric absorption (S. Walker).

Volume V is devoted to a single topic: bond lengths and angles from X-ray crystallography. This topic occupied a mere sixteen pages in Volume I; it is a measure of the immense advance that has been made in the subject that a tabulation of available molecular parameters for heterocycles fills a whole volume.

Volume VI, which is now in preparation, will consist of chapters on microwave spectroscopy, photoelectron spectroscopy, bond energies, and electron-spin resonance. Volume VI will also include comprehensive subject and author indexes to the whole treatise.

A work of this magnitude must of necessity contain many gaps and omissions. It is hoped, nevertheless, that the collection of the scattered literature references will be of considerable assistance in locating physical constants and other data.

I would like to thank the authors and Academic Press for their help and cooperation throughout the preparation of these volumes.

A. R. KATRITZKY

Contents of Other Volumes

UV Photoelectron Spectra of Heterocyclic Compounds

E. HEILBRONNER, J. P. MAIER, AND E. HASELBACH

PHYSIKALISCH-CHEMISCHES INSTITUT DER UNIVERSITÄT BASEL, SWITZERLAND

I. METHOD

The technique of photoelectron spectroscopy (PE spectroscopy) has been extensively reviewed in the past[1-20] so that we can limit ourselves only to the aspects which are relevant for this chapter. The

[1] D. W. Turner, C. Baker, A. D. Baker, and C. R. Brundle, "Molecular Photoelectron Spectroscopy," Wiley (Interscience), New York, 1970.

[2] D. W. Turner, *Advan. Phys. Org. Chem.* **4**, 31 (1966).

[3] W. C. Price, *Endeavour* **26**, 75 (1967).

[4] W. C. Price, Developments in Photo-Electron-Spectroscopy, *in* "Molecular Spectroscopy" (P. Hepple, ed.). Inst. of Petroleum, London, 1968.

[5] D. W. Turner, Molecular Photoelectron Spectroscopy, *in* "Molecular Spectroscopy" (P. Hepple, ed.). Inst. of Petroleum, London, 1968.

[6] D. W. Turner, Molecular Photoelectron Spectroscopy, *in* "Physical Methods in Advanced Inorganic Chemistry" (H. A. O. Hill and P. Day, eds.). Wiley (Interscience), New York, 1968.

[7] H. Lemaire, *Sciences* **10**, 32 (1969).

[8] D. W. Turner, *Ann. Rev. Phys. Chem.* **21**, 107 (1970).

[9] W. C. Price and D. W. Turner, *Phil. Trans. Roy. Soc. London A* **268**, 1–175 (1970).

[10] A. D. Baker, *Accounts Chem. Res.* **3**, 17 (1970).

[11] S. D. Worley, *Chem. Rev.* **71**, 295 (1971).

essential feature of the method[21,22] consists of irradiation of a (closed shell) molecule M with photons of energy $h\nu$ well in excess of the usual (first) ionization potential (IP). When the cross section for ionization of M is favorable, an electron of kinetic energy T is ejected, leaving a radical cation M^+ in an electronic state Ψ_j:

$$M + h\nu \xrightarrow{\Delta E_j} M^+(\Psi_j) + e(T) \qquad (1)$$

The energy change ΔE_j associated with the process (1) is

$$\Delta E_j = I_a(\Psi_j) + \Delta E_j(\text{vib.}) + \Delta E_j(\text{rot.}) = h\nu - T \qquad (2)$$

where $I_a(\Psi_j)$ is the adiabatic IP of M, i.e., the energy necessary to produce $M^+(\Psi_j)$ in its vibrational and rotational ground state and where M^+ assumes the equilibrium configuration and/or conformation associated with the electronic state Ψ_j. $\Delta E_j(\text{vib.})$ and $\Delta E_j(\text{rot.})$ are the energies necessary to excite $M^+(\Psi_j)$ to a particular vibrationally or rotationally excited state.

Depending on whether the photon energy $h\nu$ corresponds to the far UV region (~ 20 eV) or the X-ray region ($\sim 10^3$ eV), one differentiates between UPS (ultraviolet PE spectroscopy) and XPS (X-ray PE spectroscopy). The latter technique is also known as ESCA[23,24] (Elec-

---

[12] N. Knöpfel, Th. Olbricht and A. Schweig, *Chem. unserer Z.* **5**, 65 (1971).

[13] C. R. Brundle, *Appl. Spectrosc.* **25**, 8 (1971).

[14] C. R. Brundle and M. B. Robin, Photoelectron Spectroscopy, *in* "Determination of Organic Structures by Physical Methods" (F. C. Nachod and G. Zuckermann (ed.), Vol. 3, p. 1. Academic Press, New York, 1971.

[15] E. Heilbronner, *Int. Congr. Pure Appl. Chem., 23rd*; Suppl. *Pure Appl. Chem. Suppl.* **7**, 9 (1971).

[16] A. Hamnet and A. F. Orchard, Photoelectron Spectroscopy, *in* "Electronic Structure and Magnetism of Inorganic Compounds," Vol. 1, A Specialist Periodical Rep. Chem. Soc., London, 1972.

[17] A. D. Baker and D. Betteridge, Photoelectron Spectroscopy. Chemical and Analytical Aspects, Int. Ser. Monographs in Anal. Chem. **53** (1972).

[18] D. A. Shirley (ed.), "Electron Spectroscopy." North-Holland Publ., Amsterdam, 1972.

[19] P. Hepple (ed.) "Molecular Spectroscopy, 1971." Inst. of Petroleum, London, 1972.

[20] R. G. Albridge, Photoelectron Spectroscopy, *in* "Physical Methods in Chemistry" (A. Weissberger, ed.), Vol. 1, Part. 1. Wiley (Interscience), New York, 1972.

[21] M. I. Al-Joboury and D. W. Turner, *J. Chem. Soc.* 5141 (1963).

[22] F. I. Vilesov, B. L. Kurbatov, and A. N. Terenin, *Dokl. Acad. Nauk. USSR* **138**, 1329 (1961).

[23] K. Siegbahn, C. Nordling, A. Fahlman, R. Nordberg, K. Hamrin, J. Hedman, G. Johansson, T. Bergmark, S.-E. Karlsson, I. Lindgren, and B. Lindberg, "Atomic, Molecular and Solid State Structure by Means of Electron Spectroscopy." Almqvist and Wiksell, Stockholm, 1967.

tron Spectroscopy for Chemical Analysis). Although the primary process (1) is the same for UPS and XPS, the type of information that can be obtained from the respective techniques is quite different. In XPS, in principle, electrons ejected from the core and valence shell can be studied. However, it is appropriate to indicate that limitations, experimental and fundamental in origin, preclude the application of XPS to the type of problems considered in this chapter. The inherent linewidth of the exciting radiation ($h\nu$) is dependent on the lifetime of the excited ions emitting it, and thus in the case of the sources normally used in the two regions, e.g., AlK_α ($h\nu = 1486$ eV) and He(I) ($h\nu = 21.22$ eV) is 0.9 eV and $\sim 10^{-7}$ eV, respectively. Although the former can be narrowed to ~ 0.2 eV, utilizing monochromatization, unavoidable loss of incident flux occurs. The latter coupled with the inherent low photoionization cross section of the valence orbitals at these short wavelengths yield unfavorable signal statistics. Thus the fine detail that is readily available from UPS is largely absent. Nevertheless, when the individual valence bands can be resolved, the comparison of the relative PE band intensities of the UPS and XPS experiments can be informative. In contrast, in UPS the fundamental resolution limits are set essentially by the thermal motion of the target molecule M, and an experimental resolution of the order of 0.02 eV full width at half-height for $T = 5$ eV electrons is customary.[25] This suffices that vibrational excitation of the ion is often discernible as discrete transitions in the PE spectra. In such an instance rotational and spin–orbit broadening may contribute appreciably to the observed vibrational peak half-width.

In XPS the vacated core orbitals ψ_j are thus essentially inner shell atomic orbitals which are strongly localized in character and therefore typical for the atom in question. The spectrum is additive in character, i.e., each atom contributes a signal in a region typical for it. Consequently, this method can be of analytical value, as implied by the name ESCA. On the other hand, far UV photons eject electrons from the delocalized molecular orbitals ψ_j of the valence shell, which no longer exhibit characteristics of the individual atoms of the molecule. The spectra are therefore dependent mainly on the type of bonding and/or the topology of the molecule. Except in rare cases, the method is of rather limited analytical value, though each spectrum is unique.[17]

[24] K. Siegbahn, C. Nordling, G. Johansson, H. Hedman, P. F. Heden, K. Hamrin, U. Gelius, T. Bergmark, L. O. Werme, R. Manne, and Y. Baer, "ESCA, Applied to Free Molecules." North-Holland Publ., Amsterdam, 1969.
[25] D. W. Turner, *Proc. Roy. Soc. London A* **307**, 15 (1968).

3

It should be mentioned that the lower-lying molecular orbitals of molecules containing second row atoms provide a borderline case. These orbitals are composed predominantly from the $2s$ atomic orbitals, and the corresponding PE bands begin to exhibit the characteristic features of XPS spectra, i.e., they are located in definite energy regions, e.g., ~ 28 eV and ~ 32 eV for the $2s$ bands of nitrogen and oxygen, respectively.[26] On the other hand, appreciable atomic nature is lost through involvement in valence bonding as reflected in the IPs and by the broadness of the Franck-Condon profiles. Needless to say, such spectra can only be recorded with higher energy photons, e.g., He(II) radiation of ~ 41 eV.

In the following sections we shall discuss and interpret PE spectra of heterocyclic compounds by correlating the observed data with the results of theoretical calculations or with qualitative deductions concerning models of the valence shell of the molecule under discussion. It is of the utmost importance to realize that such a correlation is not absolute but depends heavily on the type of theoretical model chosen and also on the particular theoretical construct we wish to choose as a reference. For this reason it may be of advantage to have a closer look at the theoretical picture of the ionization process we wish to use.

To simplify matters, we disregard for the moment the contributions ΔE_j(vib.) and ΔE_j(rot.) in Eq. (2), which will not introduce major errors because of the relationship $I_a(\Psi_j) \gg \Delta E_j$(vib.) $\gg \Delta E_j$(rot.). Under these conditions, Eq. (2) simplifies to

$$I_a(\Psi_j) = h\nu - T_a \tag{3}$$

The change in kinetic energy of M^+ relative to M can be neglected.

We shall assume that the electronic ground state of the neutral molecule M is well represented by a single Slater determinant Γ of N SCF-molecular orbitals ψ_j ($j = 1, 2, \ldots, N$), each doubly occupied by electrons with antiparallel spin:

$$\Gamma = \|\psi_1\bar{\psi}_1\psi_2\bar{\psi}_2\cdots\psi_j\bar{\psi}_j\cdots\psi_N\bar{\psi}_N\| \tag{4}$$

Ejection of an electron from orbital ψ_j leads to a doublet state, which we describe by the following configurations:

$$\Psi_j^{\alpha} = \|\psi_1\bar{\psi}_1\cdots\psi_j\cdots\psi_N\bar{\psi}_N\| \tag{5}$$

$$\Psi_j^{\beta} = \|\psi_1\bar{\psi}_1\cdots\bar{\psi}_j\cdots\psi_N\bar{\psi}_N\|$$

Except in the case of large spin–orbit interactions[16] (which we shall

[26] A. W. Potts, T. A. Williams, and W. C. Price, *Faraday Discuss. Chem. Soc.* **54**, 104 (1972).

disregard) both configurations in Eq. (5) have the same energy. In particular we have

$$\Psi_0 = \|\psi_1\bar{\psi}_1 \cdots \psi_j\bar{\psi}_j \cdots \bar{\psi}_{N-1}\psi_N\| \tag{6}$$

for the electronic ground state of M^+, i.e., the state reached by removing an electron from the highest occupied molecular orbital ψ_N.

The total energy associated with Γ [Eq. (4)] (except for the nuclear–nuclear Coulomb interaction) is

$$\mathscr{E}(M) = 2 \sum_i H_{ii} + \sum_i \sum_j (2J_{ij} - K_{ij}) \tag{7}$$

where the symbols have the usual meaning.[27,28]

The energy of the radical cation $M_v^+(\Psi_j)$, taken relative to $\mathscr{E}(M)$ is

$$\mathscr{E}[M_v^+(\Psi_j)] - \mathscr{E}(M) = -\left[H_{jj} + \sum_i (2J_{ij} - K_{ij})\right] \tag{8}$$

The expression to the left in Eq. (8) corresponds to the ionization potential associated with a process (1), such that $M_v^+(\Psi_j)$ has the same geometric structure as M. Such an ionization potential is called "vertical": $I_{v,j}$. The term in square brackets on the right-hand side is by definition the orbital energy ϵ_j of ψ_j. Relation (8) was first derived by Koopmans[29,30] and is usually quoted in the form

$$I_{v,j} = -\epsilon_j \tag{9}$$

under the name of Koopmans' theorem. Note that this theorem in this simple form cannot be applied to open-shell parent molecules M.[30]

It should be realized that expression (8) can only be derived under the assumption that the orbitals ψ_i $(i = 1, 2, \ldots, N)$ are exactly the same functions of the electron coordinates x, y, z in M and in $M_v^+(\Psi_j)$. However, this is impossible. Removing an electron from orbital ψ_j changes the electron density in $M_v^+(\Psi_j)$ by $e \cdot \psi_j^2$ relative to that in M. Under the influence of this positive charge distribution the remaining $2N - 1$ electrons will rearrange, leading to a stabilization of $M_v^+(\Psi_j)$. Consequently the $I_{v,j}$ calculated according to relation (9) will be too large. A further assumption, whose consequence is more difficult to assess, is that the electron–electron correlation energy is the same in M and in $M_v^+(\Psi_j)$.

[27] L. Salem, "The Molecular Orbital Theory of Conjugated Systems." Benjamin, New York, 1966.

[28] J. N. Murrell, "The Theory of the Electronic Spectra of Organic Molecules." Methuen, London, 1963.

[29] T. Koopmans, *Physica* **1**, 104 (1934).

[30] W. G. Richards, *Int. J. Mass Spectrom. Ion Phys.* **2**, 419 (1969).

Only when these two important factors are neglected can we use relation (9) as a means of rationalizing and interpreting PE spectra. Notwithstanding these severe limitations, we shall follow the custom of applying Koopmans' theorem (9) to derive "observed" orbital energies ϵ_j from the vertical ionization potentials $I_{v,j}$ (i.e., from the PE band positions), which necessarily will contain all those corrections one would have to apply if $\mathscr{E}[M_v^+(\Psi_j')] - \mathscr{E}(M)$ were calculated properly. Experience has shown that such a procedure yields a surprisingly coherent interpretation of PE spectroscopic behavior of, for example, heterocyclic compounds.

With respect to these conventions and the limitations they imply, the principle involved in the method of PE spectroscopy can be visualized as shown schematically in Fig. 1. Irradiation of a large number of

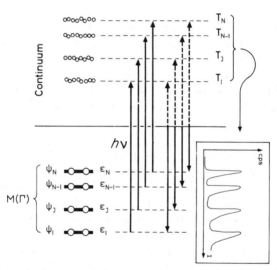

FIG. 1. Schematic representation of the principle involved in photoelectron spectroscopy, assuming the validity of Koopmans' theorem (9).

molecules M possessing a closed-shell configuration Γ of $2N$ electrons in orbitals ψ_j of energy ϵ_j, with photons of energy $h\nu$, populates the continuum with electrons of kinetic energy $T_j = h\nu - I_j$. In the PE spectrometer the continuum is scanned and the number of electrons of energy T counted.[25] The record so obtained will show peaks (in counts sec^{-1}) for $T = T_1, \ldots, T_j, \ldots, T_N$. If it is calibrated in $I = h\nu - T_j$ we obtain what can be considered a spectral representation of the orbital scheme of M.

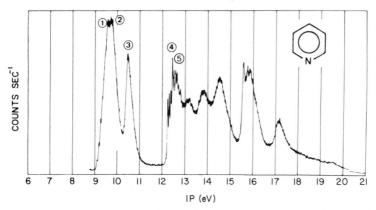

FIG. 2. PE spectrum of pyridine.

As a typical example we show in Fig. 2 the PE spectrum of pyridine.[31] An interesting feature is the integrated electron flux of each
band. Under exact experimental conditions this is related to the number
of radical ions $M^+(\Psi_j)$ in the electronic state Ψ_j. Again it is convenient
to subdivide the experimental and fundamental aspects. For a realistic
intensity comparison the electrons have to be collected over 4π radians
solid angle, or at an angle of $54°44'$ with respect to the three axes of the
coordinate system. (Two of the axes coincide with the direction of the
photon beam and the direction of maximum polarization, respectively.)
Under these conditions the relative intensities are independent of the
angular distribution of the photoelectrons and the degree of polarization
of the photon beam.[32] In all the spectra shown and in fact in those of
most of the literature covered in this review, undispersed radiation is
used and the energy analyzer employed accepts only electrons close to
$90°$ to the photon beam. An additional complication arises in the spectra
obtained with electrostatic analyzers where the effective band width is
dependent on the energy of the photoelectron (T) and this is particularly ill defined for low T. Within these limitations the partial photoionization cross sections may be compared. The latter depend on a
number of factors and these have been recently discussed in detail.[18]
However, at photon wavelengths not too near the ionization threshold,
comparison of the relative PE band intensities has been shown to
reflect orbital degeneracy (and spin degeneracy in open shell species)[33,34]

[31] R. Gleiter, E. Heilbronner, and V. Hornung, *Helv. Chim. Acta.* **55**, 255 (1972).
[32] J. A. R. Samson, *Phil. Trans. Roy. Soc. A* **268**, 141 (1970).
[33] P. A. Cox and F. A. Orchard, *Chem. Phys. Lett.* **7**, 273 (1970).
[34] P. A. Cox, S. Evans, and A. F. Orchard, *Chem. Phys. Lett.* **13**, 386 (1972).

to some extent in the consideration of similar orbitals in topologically related molecules (e.g., π bands in aromatics). In this respect the use of higher energy photon sources, e.g., He(II) ($h\nu = 40.82$ eV), is advantageous as the electron flux distribution of the resultant higher kinetic energy (T) electrons leads to less distortion in the relative intensity. Additionally, the possibility of intensity distortion due to an autoionizing process is almost certainly eliminated as few molecules will have superexcited neutral states (e.g., Rydberg states) in the 40.8 eV region.

The photoionization cross section dependence on the photon wavelength for differing symmetry characteristics of the MO's ψ_j, mentioned previously in the discussion of the intensity changes in XPS, is already reflected when He(II) radiation is used instead of the usual He(I) source.[35] This contributes to the limitation in assigning bands of PE spectra on the basis of intensities.

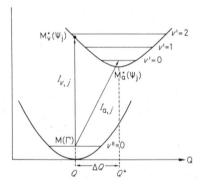

FIG. 3. Dependence of $\mathscr{E}[M(\Gamma)]$ and $\mathscr{E}[M^+(\Psi'_j)]$ on a normal coordinate Q. Q, Q^+ = position of energy minimum of M, M$^+$; $I_{v,j}$, $I_{a,j}$ = vertical, adiabatic ionization potential; v'', v' = vibrational quantum number of M, M$^+$.

Vibrational fine structure, such as discernible on the band 4 in the pyridine spectrum (Fig. 2), is quite often encountered and a brief discussion of the Franck-Condon characteristics that are recognized in connection with PE spectroscopy is necessary.[1]

In Fig. 3 is shown the semiclassical picture for the dependence of $\mathscr{E}[M(\Gamma)]$ and $\mathscr{E}[M^+(\Psi'_j)]$ on a given normal coordinate Q.

It follows from consideration of the thermal distribution of population, governed by the Boltzmann factor, exp $(-E/kT)$, that at room

[35] W. C. Price, A. W. Potts, and D. G. Streets, in "Electron Spectroscopy" (D. A. Shirley, Ed.), p. 187. North-Holland Publ., Amsterdam, 1972.

temperature ($T \approx 300°$ K) we need only to be concerned with the zeroth vibrational level $v'' = 0$ of the ground electronic state $\Gamma(M)$ of the neutral molecule M ($kT \approx 200$ cm^{-1}). Although a number of rotational levels will be significantly populated, within the experimental resolution, this will merely result in broadening of the vibrational peaks.

$M_v{}^+(\Psi'_j)$ reached "vertically" from $M(\Gamma)$ is a distorted form of $M_a{}^+(\Psi'_j)$, so that necessarily

$$\mathscr{E}[M_v{}^+(\Psi'_j)] - \mathscr{E}[M_a{}^+(\Psi'_j)] \geqslant 0 \qquad (10)$$

or

$$I_{v,j} \geqslant I_{a,j} \qquad (11)$$

In practice it is often difficult to locate $I_{v,j}$ as in general the PE bands are not symmetrical due to curvature of potential surfaces and also overlapping bands often complicate the situation. It has become customary to use the location $I_{m,j}$ of the band maximum instead of $I_{v,j}$:

$$I_{v,j} \approx I_{m,j} \qquad (12)$$

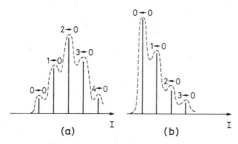

FIG. 4. Franck-Condon envelopes for (a) large ΔQ, (b) small ΔQ (cf. Fig. 3).

All vertical ionization potentials quoted in this chapter do in fact correspond to convention (12).

As suggested by Fig. 3, the transition from $M(\Gamma)$ to $M^+(\Psi'_j)$ obeys the usual Franck-Condon rules for electronic transitions.[28] Therefore the PE bands possess vibrational fine structure which can often be resolved. For example, the PE band corresponding to the situation depicted in Fig. 3 will have $2 \leftarrow 0$ as strongest component and presumably a shape as shown in Fig. 4(a). This band shape is typical for cases in which ΔQ, the relative change in position of the potential energy minimum, differs significantly from zero, i.e., if the photoelectron vacates a strongly bonding or locally antibonding orbital ψ_j. For $\Delta Q \approx 0$, i.e., if ψ_j is nonbonding, the $0 \leftarrow 0$ component will dominate and such a band will have a shape as shown in Fig. 4(b). In molecules of low symmetry, large numbers of vibrational modes can be excited and an apparent continuum band is usually obtained. Nevertheless, the profile of the Franck-Condon envelope often gives indications as discussed above.

Vibronic perturbations are pertinent in the discussion of shapes of PE bands in respect of their characteristic features which often aid in the interpretation. Evidence of spin–orbit and/or Jahn-Teller distortion in the PE spectra is associated with ionization from degenerate orbitals.[1] The former plays a prominent part in heavy atoms (e.g., Br, I) and results in an energy splitting of the electronic states corresponding to different values of total angular momentum. The effect of a Jahn-Teller type perturbation has come to be recognized in the PE spectrum as a "Jahn-Teller contour." Though the phenomenon is difficult to interpret quantitatively, the presence of such a feature suggests orbital degeneracy in the ground state. An example is afforded in the shape of the band 2 in s-triazine (Fig. 5).[31]

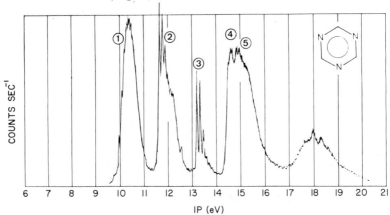

Fig. 5. PE spectrum of s-triazine.

However, a too detailed analysis of the contours of the PE bands must be treated with caution. In molecules where conformational freedom is possible, or when *in situ* decomposition can occur in the ionization chamber, the observed PE profile will result from the superposition of the individual distributions, due to the short time of the photoionization process ($\sim 10^{-16}$ sec).[36]

[36] D. L. Ames, J. P. Maier, F. Watt, and D. W. Turner, *Faraday Discuss. Chem. Soc.* **54,** 277 (1972).

II. APPLICATIONS OF PE SPECTROSCOPY

In the following sections we shall illustrate the potentialities of PE spectroscopy by discussing four typical examples taken from the field of heterocyclic compounds. These examples shall serve only as an introduction to some of the concepts which have proved to be useful, and in no way are they meant even remotely to exhaust the topic.

A. Aza Derivatives of Benzene

In Fig. 2 we have shown the PE spectrum of pyridine [2],[15,31,37–44] the first member in the series of the azaderivatives [2–7] of benzene [1].

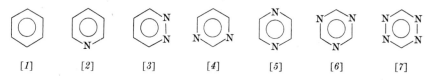

| [1] | [2] | [3] | [4] | [5] | [6] | [7] |

From a theoretical point of view, the molecules 2–7 are best regarded as perturbed benzenes, in that the replacement of a methine unit at positions ξ by a nitrogen center substitutes a carbon $2p$ atomic orbital ϕ_ξ by a nitrogen $2p$ atomic orbital ϕ_ξ^N in the linear combinations

$$\psi_j = \sum_\mu c_{j\mu}\phi_\mu \tag{13}$$

of the π system of 1. In addition, each nitrogen atom in the molecules 2–7 possesses what is usually described as an electron lone pair, i.e., two electrons occupying an atomic orbital of sp^2 character.

The first step in the interpretation of the PE spectra of the azines, e.g., the diazines pyridazine [3], pyrimidine [4] and pyrazine [5] (cf. orbital diagram Fig. 6), consists in locating those bands which correspond according to relation (1) to the ejection of an electron from a π orbital [Eq. (13)]. In particular, we are interested in the two highest

[37] M. J. S. Dewar and S. D. Worley, *J. Chem. Phys.* **51**, 263 (1969)

[38] C. Goffart, J. Momigny, and P. Natalis, *Int. J. Mass Spectrom. Ion Phys.* **3**, 37 (1969).

[39] B.-O. Jonsson, E. Lindholm, and A. Skerbele, *Int. J. Mass Spectrom. Ion Phys.* **3**, 385 (1969).

[40] A. D. Baker and D. W. Turner, *Phil. Trans. Roy. Soc. London A* **268**, 131 (1970).

[41] R. Gleiter, E. Heilbronner, and V. Hornung, *Angew. Chem. Int. Ed.* **82**, 878 (1970).

[42] E. Heilbronner, V. Hornung, F. H. Pinkerton, and S. F. Thames, *Helv. Chim. Acta* **55**, 289 (1972).

[43] G. H. King, J. N. Murrell, and R. J. Suffolk, *J. Chem. Soc. Dalton* 564 (1972).

[44] C. Batich, E. Heilbronner, V. Hornung, A. J. Ashe, III, D. T. Clark, U. T. Cobley, D. Kilcast, and I. Scanlan, *J. Amer. Chem. Soc.* **95**, 928 (1973).

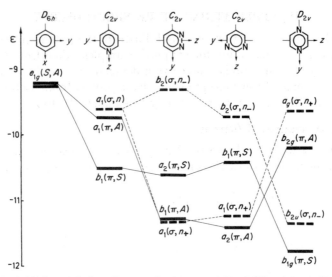

FIG. 6. Orbital correlation diagram for benzene [1], pyridine [2], pyridazine [3], pyrimidine [4], and pyrazine [5]. Solid lines = π orbital levels; broken lines = lone-pair orbital levels. S and A refer to the definition given in relation (14).

occupied π orbitals which correlate with the degenerate pair

$$e_{1g}(\pi) \begin{cases} S: (\phi_1 - \phi_4)/\sqrt{3} + (\phi_2 - \phi_3 - \phi_5 + \phi_6)/\sqrt{12} & (14) \\ A: \tfrac{1}{2}(\phi_2 + \phi_3 - \phi_5 - \phi_6) \end{cases}$$

of benzene. Previous experience with unsaturated and aromatic hydrocarbons[45-49] suggests that a simple HMO model is capable of reproducing PE π band positions with sufficient accuracy, i.e., within 0.5 eV or better.[45] The shifts $\delta\epsilon_j$ induced by the replacement of carbon by nitrogen in the orbital energies ϵ_j can be taken care of by a first-order perturbation treatment, i.e., by postulating

$$\alpha_\xi \equiv \alpha_N = \langle \phi_\xi^N | \mathscr{H} | \phi_\xi^N \rangle = \alpha_C + \delta\alpha_N \qquad (15)$$

where α_C is the Coulomb integral for the carbon atomic $2p$ orbital. Note that $\delta\alpha_N$ is a negative quantity. For the atomic orbitals ϕ_ξ and

[45] J. H. D. Eland and C. J. Danby, Z. Naturforsch. 23a, 355 (1968).

[46] J. H. D. Eland, Int. J. Mass. Spectrom. Ion Phys. 2, 471 (1969).

[47] F. Brogli and E. Heilbronner, Theoret. Chim. Acta 26, 289 (1972).

[48] F. Brogli and E. Heilbronner, Angew. Chem. 84, 551 (1972); Angew. Chem. Int. Ed. 11, 538 (1972).

[49] P. A. Clark, F. Brogli, and E. Heilbronner, Helv. Chim. Acta 55, 1415 (1972).

$\phi_{t'}$ *ortho* to the nitrogen center ξ we assume

$$\alpha_{\tau} = \alpha_{t'} = \alpha_C + m\delta\alpha_N \tag{16}$$

with $m \approx \frac{1}{3}$, to take care of the inductive influence of the electronegative nitrogen atom on the neighboring carbon and/or nitrogen atoms.[31,42,50,51] This model yields the π orbital correlation shown in the diagram of Fig. 6 (solid lines = orbital energy levels). With the parameters $\delta\alpha_N = -3.3$ eV and $m = 0.31$ [51] the quantitative agreement is almost perfect (see Table I).

Earlier attempts to interpret the PE spectra of azines[37] were hampered by the preconceived idea that lone-pair orbitals n_{ζ}, $n_{\zeta'}$

TABLE I

Compound	Parent[a] orbital	$I_{v,J}$(exp.)	$I_{v,J}$(calc.)
[1]	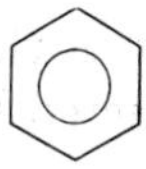 $e_{1g}(S)$ $e_{1g}(A)$	9.24	9.16
[2]	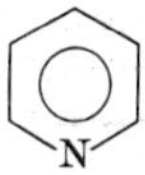 $e_{1g}(A)$ $e_{1g}(S)$	9.73 10.50	9.68 10.43
[3]	 $e_{1g}(S)$ $e_{1g}(A)$	10.61 11.30	10.57 11.32
[4]	 $e_{1g}(S)$ $e_{1g}(A)$	10.41 11.39	10.57 11.32
[5]	 $e_{1g}(A)$ $e_{1g}(S)$	10.18 11.77	10.19 11.70
[6]	 $e_{1g}(S)$ $e_{1g}(A)$	11.67	11.84
[7]	 $e_{1g}(S)$ $e_{1g}(A)$	12.10 13.59	11.98 13.49

[a] Parent orbitals (14) of unperturbed benzene used in first-order calculation.

[50] E. Heilbronner, V. Hornung, H. Bock, and H. Alt, *Angew. Chem.* **81**, 537 (1969).
[51] F. Brogli, E. Heilbronner, and T. Kobayashi, *Helv. Chim. Acta* **55**, 274 (1972).

would interact appreciably only if ζ and ζ' were neighboring centers, e.g., in pyridazine [3]. However, as the PE spectroscopic results for diazabicyclo[2.2.2]octane [8][52,53] (see Fig. 7) have shown there exists an appreciable through-bond interaction between the lone-pair orbitals

[8] n_1 ⬭N⤳N⬭ n_4

n_1 and n_4 in this compound, which yields a split of 2.2 eV between the PE bands corresponding to the molecular orbitals $\psi(n_+)$, $\psi(n_-)$ to which the linear combinations

$$n_+ = (n_1 + n_4) / \sqrt{2} \qquad (17)$$

$$n_- = (n_1 - n_4) / \sqrt{2} \qquad (18)$$

contribute with greatest weight. Contrary to the naive expectation, the $\psi(n_+)$ orbital corresponding to the first band at $I_{v,1} = 7.5$ eV in the PE spectrum of 8 lies above $\psi(n_-)$ which corresponds to the second band at $I_{v,2} = 9.7$ eV. This remarkable result was predicted by theory.[54–57]

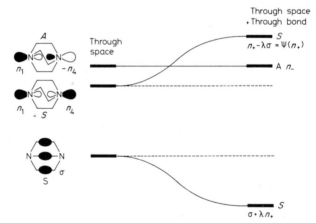

FIG. 7. Schematic orbital diagram for diazabicyclo[2.2.2]octane [8]. Left side: through-space interaction only. Toward the right: increasing through-bond interaction.

[52] P. Bischof, J. A. Hashmall, E. Heilbronner, and V. Hornung, *Tetrahedron Lett.* 4025 (1969).

[53] E. Heilbronner and K. A. Muszkat, *J. Amer. Chem. Soc.* **92**, 3818 (1970).

[54] W. Adam, *Jerusalem Symp. Quantum Chem. Biochem.*, **2**, 118. Israel Acad. Sci. and Humanities, Jerusalem, 1970.

[55] W. Adam, A. Grimison, and R. Hoffmann, *J. Amer. Chem. Soc.* **91**, 2590 (1971).

[56] R. Hoffmann, A. Imamura, and W. J. Hehre, *J. Amer. Chem. Soc.* **90**, 1499 (1968).

[57] R. Hoffmann, E. Heilbronner, and R. Gleiter, *J. Amer. Chem. Soc.* **92**, 706 (1970).

The underlying electronic mechanism was rationalized by Hoffmann in terms of a heuristically useful set of rules concerning the interplay of "through space" and "through bond" interactions between localized or semilocalized orbitals.[64]

As shown in Fig. 7 for the particular case 8, direct overlap-controlled "through space" interaction between n_1 and n_4 would yield the "natural" sequence of $n_+(S)$ below $n_-(A)$. In fact the overlap between n_1 and n_4 is rather small because of their large separation. For symmetry reasons only the lower-lying orbital n_+ of the pair (17)(18) can interact with the totally symmetric linear combination of the CC–σ orbitals. As a result the order of the orbitals is inverted with $\psi(n_+) \approx n_+ - \lambda\sigma$ above $\psi(n_-)$ (see Fig. 7).

If these rules are applied to the diazines 3, 4, and 5 the assignment of the "lone-pair"PE bands shown in the correlation diagram of Fig. 6 (dashed orbital energy levels) results.[15,31] In Fig. 8 are shown the lone-pair orbitals of 2 to 5, i.e., those molecular orbitals to which n_1 (in 2) or n_+ and n_- (in 3, 4, 5) contribute with greatest weight. It is instructive to see how in the case of pyrazine [5] "through bond" interaction mixes the CC–σ orbitals with the linear combination n_+, so that the two electrons of $a_g(\sigma; n_+)$ are much more delocalized than those of $b_{2u}(\sigma; n_-)$ which is practically devoid of CC–σ character.

The assignment given in Fig. 6 has been supported by more sophisticated calculations,[58–63] correlation with the PE spectroscopic results on triazine [6][31,65] and tetrazine [7][31,66] (see Fig. 9), correlation with the PE spectra of aza- and diazanaphthalenes,[51,67] the investigation of the

[58] I. Fischer-Hjalmars and M. Sundbom, *Acta Chem. Scand.* **22**, 607 (1968).

[59] M. Sundbom, *Jerusalem Symp. Quantum Chem. Biochem.* **2**, 56. Israel Acad. Sci. and Humanities, Jerusalem, 1970.

[60] G. Hohlneicher and W. Sänger, *Jerusalem Symp. Quantum Chem. Biochem.* Israel Acad. Sci. and Humanities, Jerusalem 1970, p. 193.

[61] L. Åsbrink, C. Fridh, B.-O. Jonsson, and E. Lindholm, *Int. J. Mass Spectrom. Ion Phys.* **8**, 229 (1972).

[62] L. Åsbrink, C. Fridh, B.-O. Jonsson, and E. Lindholm, *Int. J. Mass Spectrom. Ion Phys.* **8**, 215 (1972).

[63] C. Fridh, L. Åsbrink, B.-O. Jonsson, and E. Lindholm, *Int. J. Mass Spectrom. Ion Phys.* **8**, 101 (1972).

[64] R. Hoffmann, *Accounts Chem. Res.* **4**, 1 (1971).

[65] C. Fridh, L. Åsbrink, B.-O. Jonsson, and E. Lindholm, *Int. J. Mass Spectrom. Ion Phys.* **8**, 85 (1972).

[66] C. Fridh, L. Åsbrink, B.-O. Jonsson, and E. Lindholm, *Int. J. Mass Spectrom. Ion Phys.* **9**, 485 (1972).

[67] D. M. W. Van den Ham and D. Van der Meer, *Chem. Phys. Lett.* **12**, 447 (1972).

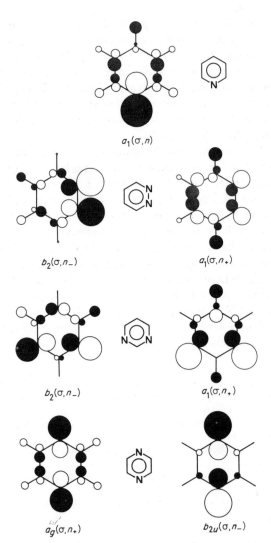

FIG. 8. Diagrams of the lone-pair orbitals $\psi(n_+)$ and $\psi(n_-)$ of pyridine [2], pyridazine [3], pyrimidine [4], and pyrazine [5]. These orbitals have been given their symmetry labels relative to C_{2v} [2, 3, 4] or D_{2h} [5].

16

dependence of PE band positions on substitution,[42,43,68,69] and finally the determination of the angular intensity dependence of the bands in the PE spectra of *1* to *4*.[36]

This assignment (see Figs. 6 and 9) is summarized in Table II (where all ionization potentials are given in electron volts).

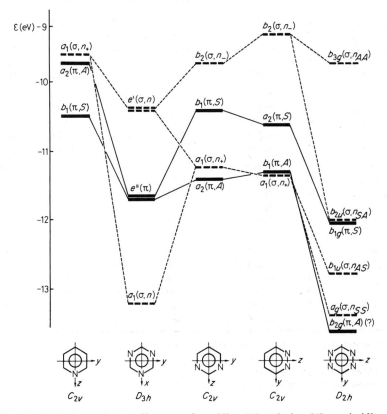

FIG. 9. Orbital correlation diagram of pyridine [*2*], triazine [*6*], pyrimidine [*4*], pyridazine [*3*], and tetrazine [*7*]. Solid lines = π orbital levels; broken lines = lone pair orbital levels. S and A for π orbitals refer to the definition given in relation (14), and n_{AA}, n_{SA}, n_{AS}, and n_{SS} of 7 refer to those linear combinations of the basis n orbitals which are symmetric (antisymmetric) relative to the x, y or x, z plane (in that order).

[68] M. B. Robin, N. A. Kuebler, and C. R. Brundle, in "Electron Spectroscopy" (D. A. Shirley, Ed.), p. 351. North-Holland Publ., Amsterdam, 1972.
[69] C. R. Brundle, M. B. Robin, and N. A. Kuebler, *J. Amer. Chem. Soc.* **94**, 1466 (1972).

TABLE II

Compound	Band	Orbital	$I_{v,J}{}^{31}$	$I_{a,J}{}^{37}$	$I_{?,J}{}^{37\,a}$	$I_{a,J}{}^{62-66}$
	1	$b_2(\sigma; n_-)$	9.31	8.64	8.90	8.71
	2	$a_2(\pi)$	10.61	10.49	10.53	10.48
	3	$b_1(\pi)$	11.3	b	11.61	11.1
	4	$a_1(\sigma; n_+)$	11.3	b	—	11.1
C_{2v}	5	$b_1(\pi)$	13.9	13.45	13.63	13.5
	1	$b_2(\sigma; n_-)$	9.73	9.23	9.42	9.32
	2	$b_1(\pi)$	10.41	10.41	10.39	10.40
	3	$a_1(\sigma; n_+)$	11.23	11.10	11.06	11.1
	4	$a_2(\pi)$	11.39	b	—	11.3
C_{2v}	5	$b_1(\pi)$	13.9	13.51	13.62	13.6
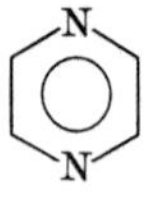	1	$a_g(\sigma; n_+)$	9.63	9.29	9.36	9.22
	2	$b_{2g}(\pi)$	10.18	10.18	10.15	10.17
	3	$b_{2u}(\sigma; n_-)$	11.35	11.00	11.14	10.93
	4	$b_{1g}(\pi)$	11.77	11.66	11.73	11.66
D_{2h}	6	$b_{3u}(\pi)$	13.9	b	13.13	13.1
	1	$e'(\sigma; n)$	10.37	9.98	—	10.01
	2	$e''(\pi)$	11.67	11.67	—	11.69
	3	$a_1'(\sigma; n)$	13.21	13.21	—	13.26
	4	$a_2''(\pi)$	14.67	14.51	—	14.56
D_{3h}	5	$e'(\sigma)$	14.88	c	—	15.00
	1	$b_{3g}(\sigma; n)$	9.72	9.14	—	—
	2	$b_{1g}(\pi)$	12.05	11.75	—	—
	3	$b_{2u}(\sigma; n)$	12.05	—	—	—
	4	$b_{1u}(\sigma; n)$	12.78	d	—	—
	5	$a_g(\sigma; n)$	13.36	d	—	—
D_{2h}	6	$b_{2g}(\pi)$	13.59	d	—	—
	7	$b_{3u}(\pi)$	15.84	15.53	—	—

a The exact nature of these ionization potentials is not defined.
b Strongly overlapping bands.
c Onset overlapped by band 4.
d Band onset is overlapped by preceding band.

B. Phosphabenzene, Arsabenzene, and Stibabenzene

Investigations of phospha and arsa derivatives of benzene[70,71] and anthracene[72] have shown that an assignment can be derived on the basis of simple perturbation considerations, similar to those embodied in formulas (15) and (16). As an extension of this work the PE spectra

[70] H. Oehling, W. Schäfer, and A. Schweig, Angew. Chem. 83, 723 (1971).
[71] A. Schweig, W. Schäfer, and K. Dimroth, Angew. Chem. 84, 636 (1972).
[72] W. Schäfer, A. Schweig, F. Bickelhaupt, and H. Vermeer, Angew. Chem. 84, 993 (1972).

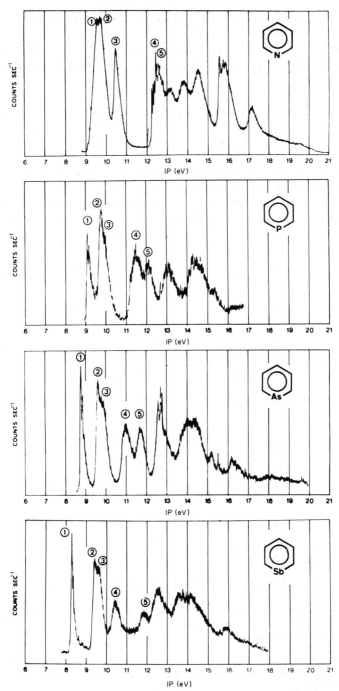

Fig. 10. PE spectra of pyridine [2], phosphabenzene [9], arsabenzene [10], and stiba-benzene [11].

19

of phosphabenzene [*9*], arsabenzene [*10*] and stibabenzene [*11*] have been recorded[44] (see Fig. 10). This series can be used to demonstrate a

correlation procedure which has also proved useful in other cases.[73,74] Taking the heterocyclic derivatives *2, 9, 10,* and *11* of benzene as an example, the method consists in plotting ionization potentials corresponding to the ejection [relation (1)] of an electron from those π

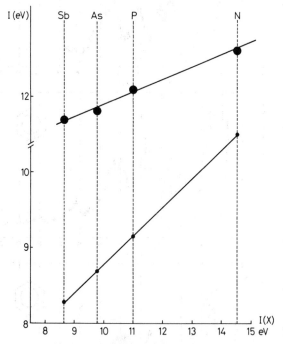

Fig. 11. Regression of the ionization potentials corresponding to electron ejection from the π orbitals of B_1 symmetry of *2, 9, 10,* and *11* on the ionization potentials $I(X)$ of the free atoms N, P, As, and Sb.

[73] D. C. Frost, C. A. McDowell, and D. A. Vroom, *J. Chem. Phys.* **46,** 4255 (1967).
[74] H. J. Haink, E. Heilbronner, V. Hornung, and E. Kloster-Jensen, *Helv. Chim. Acta* **53,** 1073 (1970).

orbitals which are derived from orbitals $e_{1g}(\pi)$, S [see (14)] or

$$a_{2u}(\pi) = (\phi_1 + \phi_2 + \phi_3 + \phi_4 + \phi_5 + \phi_6)/\sqrt{6} \qquad (19)$$

of benzene against the first ionization potential of the free nitrogen, phosphorus, arsenic, and antimony atoms.[75] Note that $e_{1g}(\pi)$, S and $a_{2u}(\pi)$ are those π orbitals of 1 which possess finite atomic orbital coefficients in position 1, i.e., in the position which is occupied by the hetero atom in 2, 9, 10, and 11. In Fig. 11 are shown the regression lines so obtained. The slopes of 0.36 for the upper $b_1(\pi)$ orbitals and of 0.16 for the second $b_1(\pi)$ orbitals agree closely with the values $\frac{1}{3}$ and $\frac{1}{6}$ predicted from the unperturbed orbitals $e_{1g}(\pi)$ (14) and $a_{2u}(\pi)$ (19) of 1. On the basis of such an empirical correlation it is now possible to establish the orbital assignment shown in Fig. 12 and summarized in Table III.

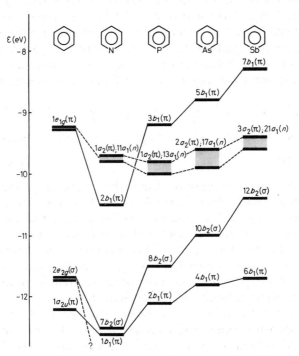

FIG. 12. Orbital correlation diagram of benzene [1], pyridine [2], phosphabenzene [9], arsabenzene [10], and stibabenzene [11]. The relative sequence of the π and n orbitals (of A_2 and A_1 symmetry, respectively) is uncertain.

[75] C. E. Moore(-Sitterley), *Nat. Bur. Std. U.S. Circ. 467* **1** (1949); **2** (1952); **3** (1958).

TABLE III

Band	[2] Orbital	I_v(eV)	[9] Orbital	I_v(eV)	[10] Orbital	I_v(eV)	[11] Orbital	I_v(eV)
1	$11a_1(\sigma; n)$	9.7	$3b_1(\pi)$	9.2	$5b_1(\pi)$	8.8	$7b_1(\pi)$	8.3
2	$1a_2(\pi)$	9.8	$1a_2(\pi)$	9.8	$2a_2(\pi)$	9.6	$3a_2(\pi)$	9.4
3	$2b_1(\pi)$	10.5	$13a_1(\sigma; n)$	10.0	$17a_1(\sigma; n)$	9.9	$21a_1(\sigma; n)$	9.6
4	$7b_2(\sigma)$	12.5	$8b_2(\sigma)$	11.5	$10b_2(\sigma)$	11.0	$12b_2(\sigma)$	10.4
5	$1b_1(\pi)$	12.6	$2b_1(\pi)$	12.1	$4b_1(\pi)$	11.8	$6b_1(\pi)$	11.7

C. Lone-Pair Interactions in Saturated Heterocyclics

The discussion of the interaction of semilocalized orbitals, e.g., "lone pairs," in molecules in terms of "through space" and "through bond" mechanisms,[64] as exemplified by 8, affords a rationalization of the ionization potentials of the lone pairs in the PE spectra of saturated oxygen and sulfur heterocycles,[76,77] e.g., in 1,2-dithiane [12],[77] 1,3-dithiane [13], and 1,4-dithiane [14].[76,77] These molecules are particularly

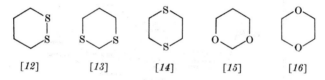

[12]   [13]   [14]   [15]   [16]

expedient for two main reasons. First, the ionization potentials of the band associated with photoelectrons ejected from the lone-pair orbitals of the oxygen and/or sulfur atoms are well separated from the remaining σ bands, and second, the Franck-Condon profiles are sharp as shown for example for 13 and 14 in Fig. 13. This is a consequence of favorable local symmetry conditions which demand that the lone-pair orbitals are essentially of pure p type, in contrast to the situation in nitrogen heterocycles where these orbitals are sp^3 in character.

At this point it may be appropriate to consider briefly what we wish to call a lone-pair orbital and to relate this to the concept of a non-bonding orbital from the point of view of PE spectroscopy. In Fig. 4(b) we have drawn the Franck-Condon profile with dominant $0 \leftarrow 0$ transition that is characteristic for the PE band associated with the

[76] D. A. Sweigart and D. W. Turner, J. Amer. Chem. Soc. 94, 5599 (1972).
[77] H. Bock and G. Wagner, Angew. Chem. 84, 119 (1972).

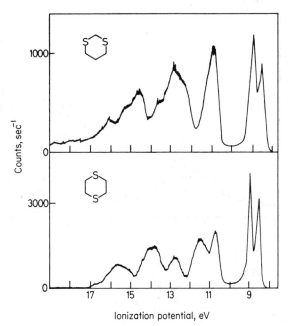

FIG. 13. PE spectra of 1,3-dithiane [13] and 1,4-dithiane [14]. Redrawn from D. A. Sweigart and D. W. Turner, *J. Amer. Chem. Soc.* **94**, 5599 (1972), and H. Bock and G. Wagner, *Angew. Chem.* **84**, 119 (1972).

ejection of an electron from a nonbonding (or weakly bonding) orbital ψ_j, i.e., for a process (1) where the potential surfaces of M and $M^+(\Psi'_j)$ are only a little different. Some lone-pair orbitals fall into this category, e.g., the highest occupied orbital of water [see Fig. 14(a)].[1] In this case the electron is ejected from a pure p-type orbital, and the Franck-Condon profile in the PE spectrum implies that the potential surfaces of the ground $(^1A_1)$ and ionic $(^2B_1)$ states are very similar, i.e., that process (1) is accompanied by very little change in geometry.[78] In saturated oxygen and sulfur derivatives, though the local symmetry is lower, a similar situation is encountered. According to the criterion mentioned above, the lone-pair orbital n_{CN} of the cyano group must also be classified as nonbonding, as can be seen from the PE spectrum of, e.g., cyanobromide[79] BrCN [Fig. 14(b)] in which the second band at 13.6 eV correspond to the ejection of an electron from n_{CN}. In contrast, the PE bands associated with the lone pair of ammonia[1,80] [Fig. 12(c)]

[78] C. R. Brundle and D. W. Turner, *Proc. Roy. Soc. London A* **307**, 27 (1968).

[79] E. Heilbronner, V. Hornung, and K. A. Muszkat, *Helv. Chim. Acta* **53**, 347 (1970).

[80] A. W. Potts and W. C. Price, *Proc. Roy. Soc. London A* **326**, 181 (1972).

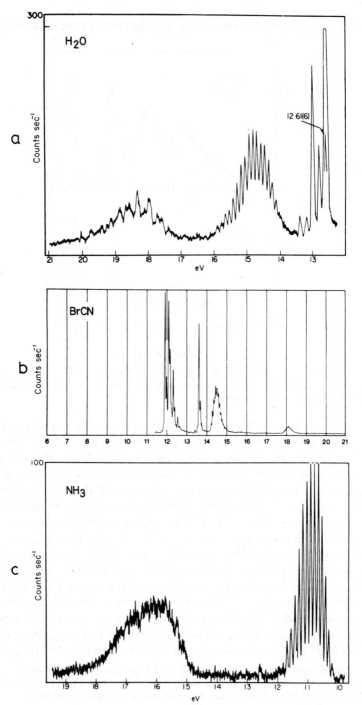

FIG. 14. PE spectra of water, cyanobromide and ammonia. Redrawn from D. W. Turner, C. Baker, A. D. Baker, and C. R. Brundle, "Molecular Photoelectron Spectroscopy." Wiley (Interscience), New York, 1970, E. Heilbronner, V. Hornung, and K. A. Muszkat, *Helv. Chim. Acta* **53**, 347 (1970), and A. W. Potts and W. C. Price, *Proc. Roy. Soc. London A* **326**, 181 (1972).

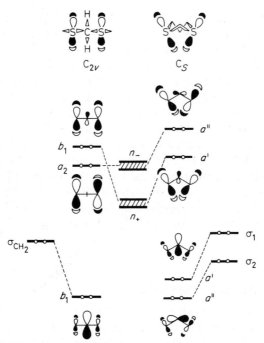

FIG. 15. Diagrammatic representation of through-space and through-bond interaction between the sulfur $3p$ lone pair orbitals in 1,3-dithiane [*13*]. Redrawn from H. Bock and G. Wagner, *Angew. Chem.* **84**, 119 (1972).

is drastically different. The orbital associated with the lone pair can mix with other, deeper orbitals of the same symmetry and thus through delocalization acquire "bonding" characteristics. As a consequence, ejection of an electron from this orbital leads to a large geometry change, and in fact the ionic state $(^2A_1)$ of M^+ is planar while the ground state $(^1A_1)$ of M is pyramidal. Clearly the lone-pair PE band of ammonia at 11 eV [Fig. 14(c)] is of the type depicted in Fig. 4(a).

In *12* the direct overlap between the sulfur lone-pair orbitals n_1 and n_2 determines their interaction.[77] The difference in the ionization potentials corresponding to ejection from the symmetric (n_+) and antisymmetric (n_-) linear combinations [see relations (17,18)] is 0.95 eV, as expected for two bonded p-type n orbitals twisted by 60° with respect to each other. In contrast in *14* and in 1,4-dioxane [*16*] the splittings of 0.45 and 1.22 eV, respectively,[76] cannot be attributed to through-space interaction alone (e.g., distance $S\cdots S \approx 3.5$ Å) but are mainly due to through-bond coupling. Significantly, the splitting is greater when the heteroatoms are oxygen, in accord with the postulated mechanism. The more tightly bound oxygen lone-pair orbitals provide a more favorable energy match with the σ framework than do the sulfur ones.

An intermediate situation is provided by *13* and by 1,3-dioxane [*15*][76,77] where both through-space and through-bond interactions occur simultaneously, yielding a splitting of 0.41 and 0.25 eV, respectively. An interesting facet of the through-bond interaction is illustrated in these latter molecules. The natural sequence of n_+ below n_- will be retained even under conditions of strong through-bond coupling, when interaction with the more tightly bound sigma orbitals is allowed for both combinations of the lone pairs and if the orbitals obey the natural sequence. This is shown diagrammatically in Fig. 15. That the splitting is greater in *13* than *15* suggests that through-space interaction is more important in the former due to the greater size of the sulfur lone-pair orbitals.

D. Determination of Conformations

Usually prior knowledge of the equilibrium conformation of the parent molecule M in its electronic ground state is necessary for the interpretation of its PE spectrum. Only if certain PE bands or groups of bands are very sensitive to conformational changes in a predictable way can PE spectroscopy be used safely for the determination and analysis of such changes.

Typical examples are molecules RS in which two π systems R and S are joined by an essential single bond, e.g., two ethylene moieties in butadiene. In this case the compound molecule RS will exhibit π orbitals which can be written in a first approximation as linear combinations of the semilocalized π orbitals ψ_R, ψ_S of the parent systems: R and S.

$$\psi_{RS,+} = r\psi_R + s\psi_S \tag{20}$$

$$\psi_{RS,-} = s\psi_R - r\psi_S$$

The difference ΔI between the ionization potentials associated with the ejection of an electron from either $\psi_{RS,+}$ or $\psi_{RS,-}$ depends on the amount of coupling $B_{RS} = \langle \psi_R | \mathscr{H} | \psi_S \rangle$ between ψ_R and ψ_S in RS, which, in turn, depends roughly on the cosine of the twist angle θ between the planes of the π systems in R and S, i.e.,

$$|\Delta I| = |\epsilon(\psi_{RS,+}) - \epsilon(\psi_{RS,-})| \propto \cos \theta \tag{21}$$

The usefulness of formula (21) has been demonstrated by the PE spectroscopic analysis of the conformations of, e.g., substituted butadienes,[81] anilines,[82,83] and nitrobenzenes[84] or styrenes[85] and biphenyls.[86,87]

An interesting case is that of the pair 2,2'-bipyridyl [*17*] and 4,4'-bipyridyl [*18*]. Application of formula (21), which had been previously calibrated on a series of substituted biphenyls, yields the result that the

[*17*]   [*18*]

twist angle θ is approximately 45° in *18*, i.e., close to the value observed for biphenyl in the gas phase, whereas *17* is planar.[86]

Another example is provided by the PE spectroscopic investigation of diaziridines [*19*]. The observed splits between the bands associated with those orbitals $\psi(n_+)$, $\psi(n_-)$, which are mainly of type (17,18), indi-

[*19*]

R, R' = H, H; H, CH$_3$; CH$_3$, CH$_3$

cate that the lone-pair orbitals n_1, n_2 and thus the substituents R, R' of *19* prefer the *trans* conformation.[88] *19* (H, H) is one of the rare cases where PE spectroscopy is superior to other instrumental techniques for reaching such a conclusion.

Under favorable conditions it is sometimes possible to analyze conformational equilibria. If two or more conformers are present in the ionization region, the PE spectrum obtained will be the superposition of the spectra of the individual conformers. This technique has been applied to the study of the conformations and conformational equilibria of substituted hydrazines.[89-91] For example, the PE spectrum of 1,2-

[81] C. R. Brundle and M. B. Robin, *J. Amer. Chem. Soc.* **92**, 5550 (1970).

[82] J. P. Maier and D. W. Turner, *Faraday Trans. II* 521 (1973).

[83] T. Kobayashi and S. Nagakura, *Chem. Lett (Chem. Soc. Japan)* 1013 (1972).

[84] T. Kobayashi and S. Nagakura, *Chem. Lett. (Chem. Soc. Japan)* 903 (1972).

[85] J. P. Maier and D. W. Turner, *Faraday Trans. II* 196 (1973).

[86] J. P. Maier and D. W. Turner, *Faraday Discuss. Chem. Soc.* **54**, 149 (1972).

[87] J. Daintith, J. P. Maier, D. A. Sweigart, and D. W. Turner, *in* "Electron Spectroscopy" (D. A. Shirley, ed.), p. 289. North-Holland Publ., Amsterdam, 1972.

[88] E. Haselbach, A. Mannschreck, and W. Seitz, *Helv. Chim. Acta* **56**, 1614 (1973).

[89] S. F. Nelsen and J. M. Buschek, *J. Amer. Chem. Soc.* **95**, 2011 (1973).

[90] S. F. Nelsen, J. M. Buschek, and P. J. Hinz, *J. Amer. Chem. Soc.* **95**, 2013 (1973).

[91] P. Rademacher, *Angew. Chem.* **85**, 447 (1973); *Angew. Chem. Int. Ed.* **12**, 408 (1973).

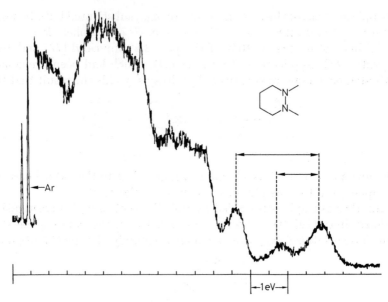

FIG. 16. PE spectrum of 1,2-dimethylhexahydropyridazine [20]. Redrawn from S. F. Nelson, J. M. Buschek, and P. J. Hinz, *J. Amer. Chem. Soc.* **95**, 2013 (1973).

dimethylhexahydropyridazine [20] shown in Fig. 16 exhibits four lone-pair bands, of which the first two overlap. The analysis of the spectrum

[20]

[20a]   [20b]   [20c]

reveals that at least two conformers of *20* must be present. The dominating conformation is the one with biaxial lone-pairs *20a*, which gives rise to the two bands showing the larger split. The two minor components yielding the small split in the PE spectrum are then attributed to *20b* and/or *20c*.[90]

28

E. Correlation between Ionization Potentials and Excitation Energies

The following section is devoted to the question of whether the results obtained from PE spectroscopy can lead to a better understanding of electronically excited states. It is known that the numerical agreement between theoretical model calculations and experimental data is generally more satisfactory for ionization than for electronic excitation processes. The reason is that the former process can be discussed reasonably well in terms of occupied orbitals only [Koopmans' theorem (9)] whereas the latter process needs in addition the virtual orbitals and orbital energies. Consequently, the ionic states $M^+(\Psi_j)$ observed in the PE experiment can usually be described by single configurations [e.g., relations (5) and (6)]. Only in exceptional cases has one to consider configurations where an electron has been removed from an orbital ψ_j and a second electron is simultaneously excited from another orbital ψ_i to an antibonding orbital ψ_k^*. Since transitions from the neutral ground state Γ of M [cf. relation (4)] to such an ionic state are strictly forbidden, the appearance of corresponding PE bands suggests that they borrow intensity from "Koopmans allowed" configurations $M^+(\Psi_j)$. The theoretical description of such states requires configuration interaction.[92]

For the description of electronically excited states of M, a configuration interaction treatment is necessary. For this reason a direct correlation of ionization potentials with electronic excitation energies is usually not feasible, except in cases in which low-lying excited states can be reasonably well described by single configurations. This condition will most probably be met if (a) the molecule has high symmetry, leading to factorization of the CI-matrix (configuration interaction) into submatrices of different symmetry; and if (b) within one submatrix the individual configurations have rather different energies and thus interact only weakly.

Thus, (b) will not be satisfied, e.g., in cases of orbital degeneracy or of the configurational degeneracy in alternant systems. It should hold for transitions such as $n \to \pi^*$ in compounds with heteroatoms, the n orbital being formally a σ orbital but separated from the lower lying σ orbital by a considerable energy gap.

Satisfactory correlations between the lone-pair ionization potentials and $n \to \pi^*$ excitation energies has been established for *trans*-azomethane[93] and *p*-benzoquinone.[94] The correlation procedure will now

[92] J. C. Lorquet and C. Cadet, *Int. J. Mass Spectrom. Ion Phys.* 7, 245 (1971).
[93] E. Haselbach and A. Schmelzer, *Helv. Chim. Acta* 54, 1575 (1971).
[94] E. Haselbach and A. Schmelzer, *Helv. Chim. Acta* 55, 1745, 3130 (1972).

be illustrated, using pyrazine [5] and 2,6-dimethylpyrazine [21] as an example.[95]

[21]

In 5 the splitting between the lone-pair PE bands (see Fig. 6) is

$$\Delta I = I(n_-) - I(n_+) = \epsilon[\psi(n_+)] - \epsilon[\psi(n_-)] = 1.72 \text{ eV} \qquad (22)$$

Simple orbital diagrams such as the one given in Fig. 17 suggest that

$$\Delta E = E(n_- \to \pi^*) - E(n_+ \to \pi^*) \approx \Delta I \qquad (23)$$

Such a view seems to be supported by theoretical calculations on 5 which yield 1.82 eV,[96] 1.4 eV,[97] or 1.3 eV[98] for ΔE.[99] On the other hand, several experimental studies indicate that ΔE for 5 and 21 is rather small, having a value not larger than 0.2 eV.[100-106]

This discrepancy can be resolved by considering the expression for an excited singlet configuration in the framework of an SCF MO model. This yields the expectation value

$$E(\psi_m \to \psi_n) = \epsilon_n - \epsilon_m - J_{mn} + 2K_{mn} \qquad (24)$$

For $n \to \pi^*$ transitions, K_{mn} turns out to be very small and can be neglected. Furthermore, $\epsilon_m = -I(\psi_m)$ according to Koopmans' theorem (9). Using the abreviations defined in relations (22) and (23) the splitting ΔE between the $n \to \pi^*$ states in 5 is given by

$$\Delta E \approx \Delta I + J_{n_+\pi^*} - J_{n_-\pi^*} = \Delta I + \Delta J \qquad (25)$$

Note that by taking the difference between the two $n \to \pi^*$ transitions,

[95] E. Haselbach, Z. Lanyiova, and M. Rossi, Helv. Chim. Acta 56, 2889 (1973).
[96] T. Yonezawa, H. Kato, and H. Kato, Theoret. Chim. Acta 13, 125 (1969).
[97] M. Hackmeyer and J. W. Whitten, J. Chem. Phys. 54, 3739 (1971).
[98] R. L. Ellis, G. Kuehnlenz, and H. H. Jaffe, Theoret. Chim. Acta 26, 131 (1972).
[99] A. D. Jordan, I. G. Ross, R. Hoffmann, I. R. Swenson, and R. Gleiter, Chem. Phys. Lett. 10, 572 (1971).
[100] G. W. Robinson and M. A. El-Sayed, Mol. Phys. 4, 273 (1961).
[101] R. Hochstrasser and C. Marzzacco, J. Chem. Phys. 49, 971 (1968).
[102] Y. H. Li and E. C. Lim, Chem. Phys. Lett. 9, 514 (1971).
[103] C. J. Marzzacco and E. F. Zalewski, J. Mol. Spectrosc. 43, 239 (1972).
[104] K. K. Innes, A. H. Kalantar, A. Y. Khan, and T. J. Durnick, J. Mol. Spectrosc. 43, 477 (1972).
[105] W. R. Moomaw, M. R. Decamp, and P. C. Podore, Chem. Phys. Lett. 14, 255 (1972).
[106] K. K. Innes, J. P. Byrne, and I. G. Ross, J. Mol. Spectrosc. 22, 125 (1967).

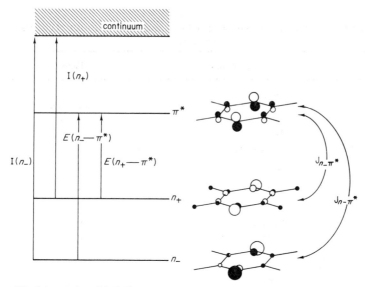

FIG. 17. Schematic orbital diagram for the comparison of ionization potentials and electronic excitation energies in pyrazine [5].

the virtual orbital energy ϵ_{π^*} cancels. Obviously, relation (23) would hold if $\Delta J = 0$, i.e., if the repulsion between the n_+ electron and the π^* electron equals the repulsion between the n_- electron and the π^* electron. From the diagramatic representations of these molecular orbitals given in Fig. 17 it is evident that this is not the case. Thus, the n_+ MO has less space in common with the π^* MO than the n_- MO so that $J_{n_+\pi^*} < J_{n_-\pi^*}$, or $\Delta J < 0$. According to relation (25), this yields $\Delta E < \Delta I$, in qualitative agreement with the experimental evidence. Explicit calculation of ΔJ using the CNDO/2 method[107] yields $\Delta J = -1.06$ eV for 5 and $\Delta J = -1.27$ eV for 21. From PE spectroscopy, $\Delta I = 1.72$ eV for 5 and $\Delta I = 1.54$ eV for 21. This leads according to relation (25) to $\Delta E = 0.66$ eV for 5 and $\Delta E = 0.27$ eV for 21. Although there is no perfect numerical agreement with the experimental evidence for ΔE, the above results definitely support the idea that the splitting ΔE between the two $n \rightarrow \pi^*$ states in 5 and 21 is much smaller than the splitting ΔI. The reason for this discrepancy is that the two lone-pair orbitals $\psi(n_-)$ and $\psi(n_+)$ have different shapes. As has been discussed

---

[107] J. Pople and D. L. Beveridge, "Approximate MO Theory." McGraw Hill, New York, 1970.

TABLE IV

HETEROCYCLIC COMPOUNDS WHOSE PHOTOELECTRON SPECTRA HAVE BEEN DETERMINED[a]

R	H	N	Compound	Reference[b]
1	1	3	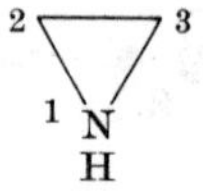	108, 109
			2,2-Dimethyl	109
				108, 110, 164
			2-CH_2F	110
			2-CH_2Cl	110
			2-CH_2Br	110
				111, 164
				112
1	1	4	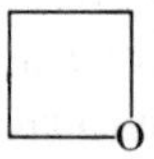	161
1	1	5	Heteroatom: nitrogen	
			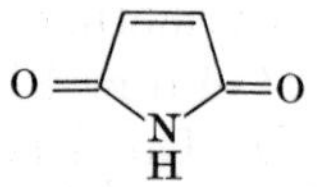	113
			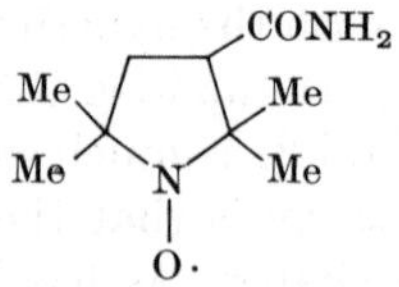	114

(continued)

[a] In view of the heterogeneous character of the list, the compounds have been ordered according to three indices: R = number of rings, H = number of heteroatoms, and N = size of ring(s).

TABLE IV (continued)

R	H	N	Compound	Reference[b]
1	1	5	Heteroatom: nitrogen	

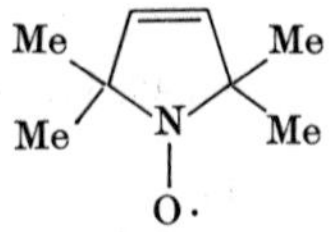

114

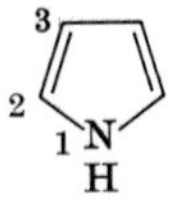

46, 115, 116, 150

3-Methyl

116

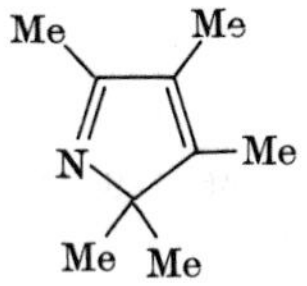

117

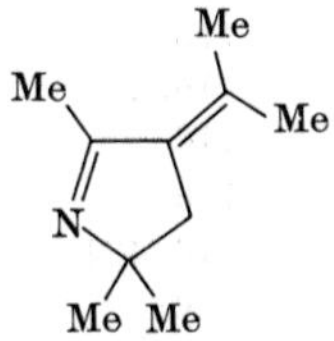

117

Heteroatom: oxygen

118

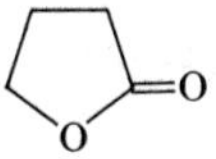

165

118

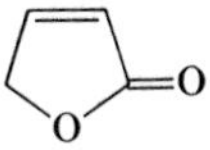

165

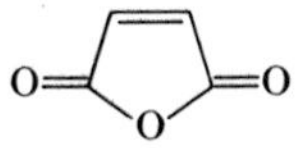

113

(continued)

33

TABLE IV (continued)

R	H	N	Compound	Reference[b]
1	1	5	Heteroatom: oxygen	

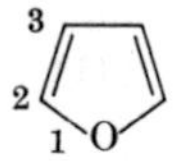	40, 46, 115, 116, 119, 120
2-CHO	40
2-Cl	40, 120

Heteroatom: sulfur

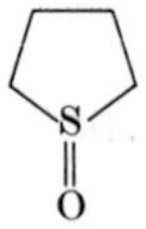

112

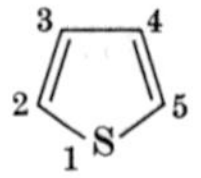

46, 116, 119, 121, 122, 162

2-Methyl	116
2-Cl	116
2-Br	116, 121
2-I	116
3-Methyl	116
3-Br	116, 121
2,5-DiCl	116

Other heteroatoms

162, 163

162, 163

123

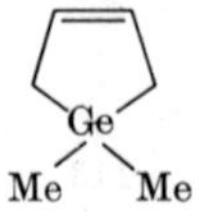

123

(continued)

TABLE IV (*continued*)

R	H	N	Compound	Reference[b]
1	1	5	Other heteroatoms	
			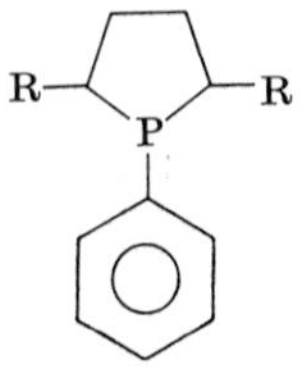 R = H, Me	154
			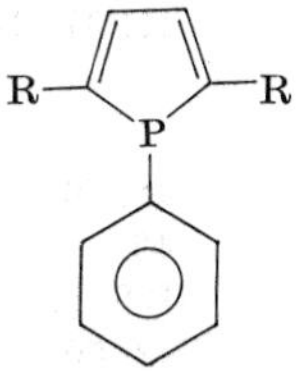 R = H, Me	154
			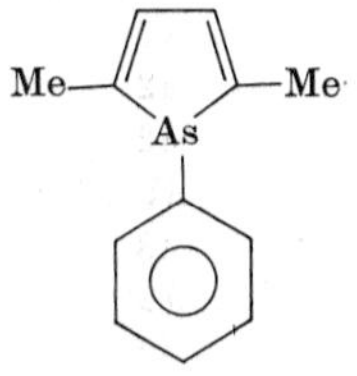	154
1	1	6	Heteroatom: nitrogen	
			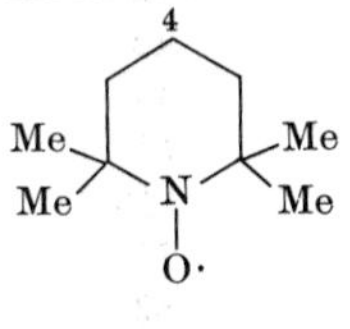	124
			4-$NHCOCH_3$	114
			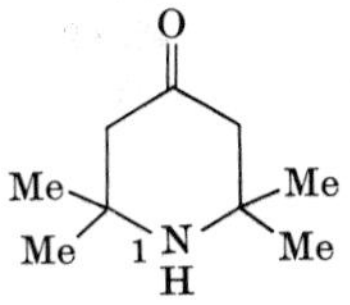	114
			1-OH	114

(*continued*)

TABLE IV (*continued*)

R	H	N	Compound	Reference[b]
1	1	6	Heteroatom: nitrogen	

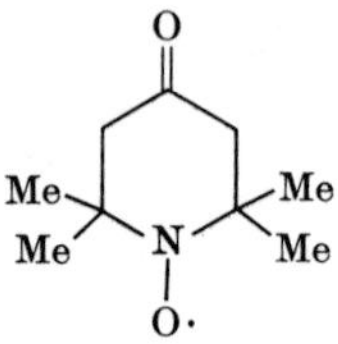

114

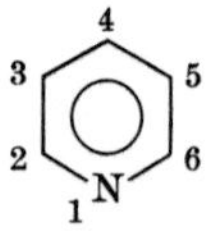

15, 31, 37–44, 69

2-Methyl	42, 50
2-F	43
2-Cl	120, 125
2-Si(CH$_3$)$_3$	42, 50
3-F	43
3-Cl	120, 125
4-Methyl	42
4-t-Butyl	42
4-Si(CH$_3$)$_3$	42
4-Cl	120, 125
2,4-Di F	43
2,4-Di Cl	125
2,5-Dimethyl	42
2,5-Di Si(CH$_3$)$_3$	42
2,6-Dimethyl	42
2,6-Di Si(CH$_3$)$_3$	42
2,6-Di F	43
2,6-Di Cl	125
3,5-Di Cl	125
2,4,6-Tri-t-butyl	70
2,4,6-Tri F	43
2,4,6-Tri Cl	125
2,3,4,5-Tetra Cl	125
Penta-F	43, 69
Penta-Cl	125

Heteroatom: oxygen

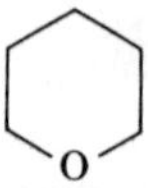

76, 77, 158

(*continued*)

36

TABLE IV (*continued*)

R	H	N	Compound	Reference[b]
1	1	6	Heteroatom: sulfur	
			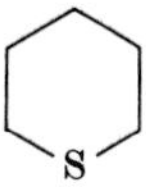	76, 126
			Other heteroatoms	
			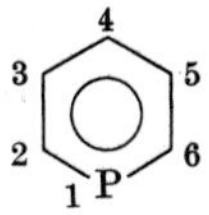	44
			2,4,6-Tri-*t*-butyl	70, 71
			1,1-Dimethyl-2,4,6-tri-*t*-butyl	71
				44
				44
1	2	3	Heteroatoms: nitrogen, nitrogen	
			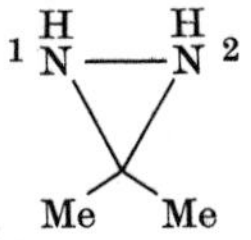	88
			1-Methyl	88
			1,2-Dimethyl	88
			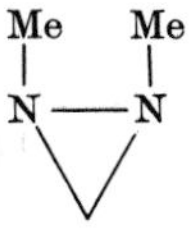	88
			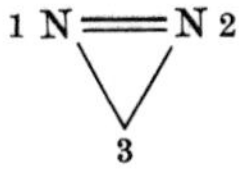	127
			3,3-Dimethyl	128
			3,3-Di F	127, 129

(*continued*)

TABLE IV (*continued*)

R	H	N	Compound	Reference[b]
1	2	5	Heteroatoms: nitrogen, nitrogen 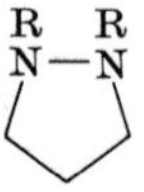	

R = H · 91

i-Pr · 89

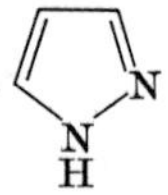

150

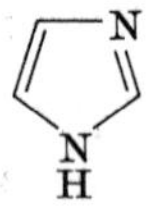

150

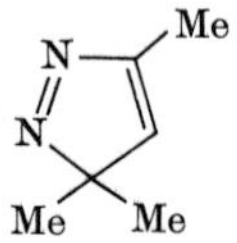

117

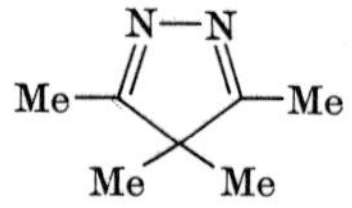

91, 117

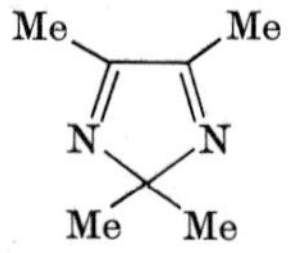

117

Heteroatoms: nitrogen, oxygen

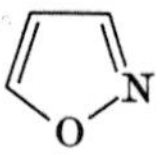

115, 116

Heteroatoms: oxygen, oxygen

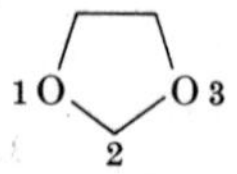

76

2,2-Dimethyl · · · · · · · · · · · · · · · · · 76

(*continued*)

TABLE IV (*continued*)

R	H	N	Compound	Reference[b]
1	2	5	Heteroatoms: oxygen, oxygen	

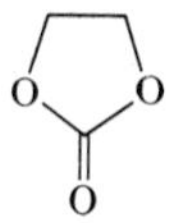

126

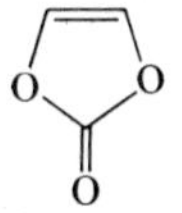

165

Heteroatoms: sulfur, sulfur

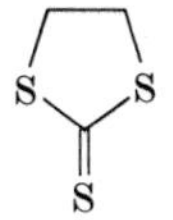

126

| 1 | 2 | 6 | Heteroatoms: nitrogen, nitrogen |

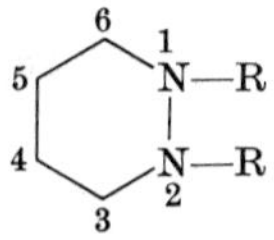

R = H	91
R = Me	90
3-methyl	90
3,6-*trans*-Dimethyl	90
3,6-*cis*-Dimethyl	90

15, 31, 36, 37, 40, 41, 62

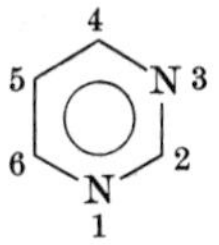

15, 31, 36, 37, 40, 41, 63

2-NH$_2$	130
2-OCH$_3$	130

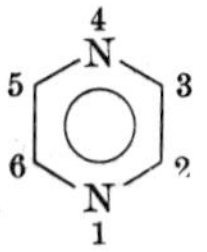

15, 31, 36, 37, 40, 41, 46, 64, 95

(*continued*)

TABLE IV (*continued*)

R	H	N	Compound	Reference[b]
1	2	6	Heteroatoms: nitrogen, nitrogen	
			2,6-Dimethyl	95
			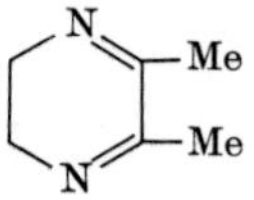	117
			Heteroatoms: oxygen, oxygen	
			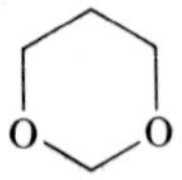	76, 158
			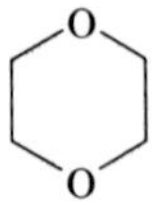	76, 158
			Heteroatom: oxygen, sulfur	
			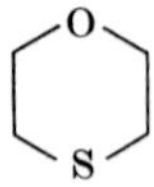	76, 126
			Heteroatoms: sulfur, sulfur	
			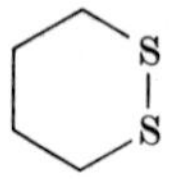	77, 153
				76, 77
			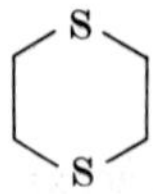	76, 77
1	3	5	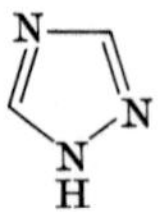	150
			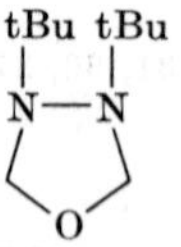	89

(*continued*)

TABLE IV (*continued*)

R	H	N	Compound	Reference[b]
1	3	5	Heteroatoms: sulfur, sulfur	
			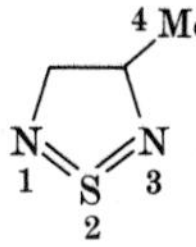	131
			4,4-Dimethyl	131
			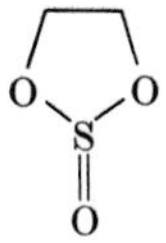	112
			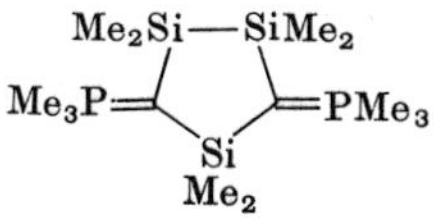	152
1	3	6	Heteroatoms: nitrogen	
			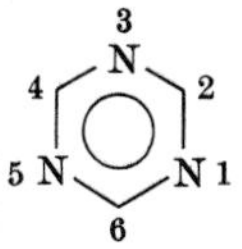	15, 31, 65, 68, 69
			2,4,6-Tri F	68, 69
			Heteroatoms: oxygen	
			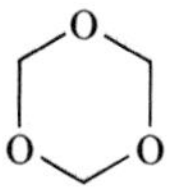	76
			Heteroatoms: sulfur	
			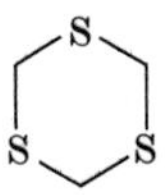	76
1	4	5	Heteroatoms: nitrogen	
			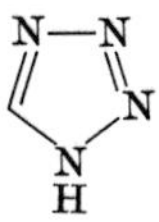	150

(*continued*)

41

TABLE IV (*continued*)

R	H	N	Compound	Reference[b]
1	4	6	Heteroatoms: nitrogen	

			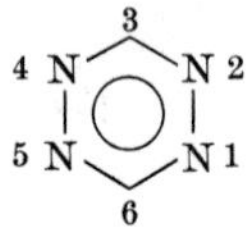	15, 31, 66
			3,6-Dimethyl	31
1	6	6	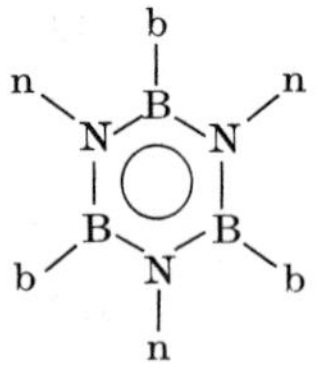	

b	n	
H	H	132–136
H	Methyl	135, 136
Methyl	H	135, 136
Methyl	Methyl	135, 136
F	H	132, 135, 136
F	Methyl	135, 136
Cl	H	135, 136
Cl	Methyl	135, 136

R	H	N	Compound	Reference
2	1	5; 6	Heteroatom: nitrogen	

			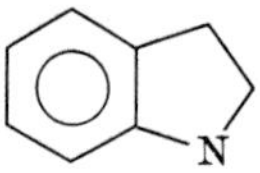	82

			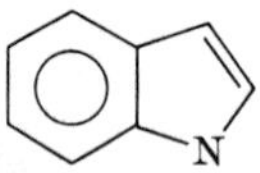	46

Heteroatom: oxygen

			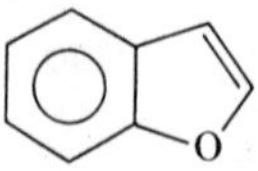	46

(*continued*)

TABLE IV (*continued*)

R	H	N	Compound	Reference[b]
2	1	5; 6	Heteroatom: sulfur	
			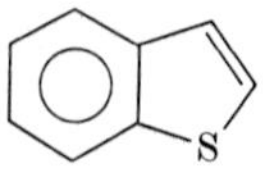	46, 156
			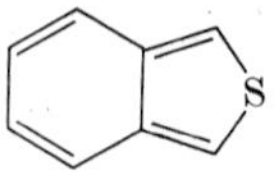	137
2	1	6; 6	Heteroatom: nitrogen	
			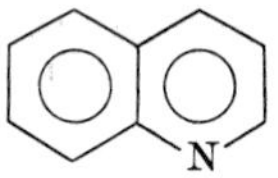	15, 37, 51, 138, 139, 157
			Hepta F	138
			2-Phenyl	157
			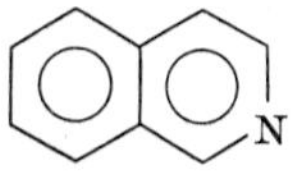	15, 37, 51, 138, 139, 157
			Hepta F	138
			Heteroatom: phosphorus	
			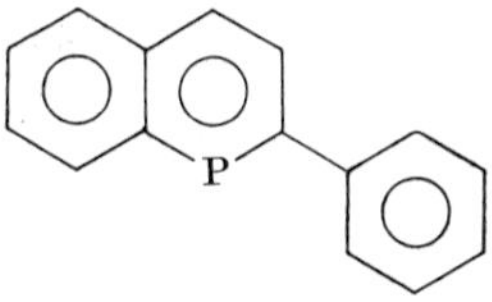	157
			1, 1-Dibenzyl	157
2	2	3; 3	Heteroatoms; nitrogen	
			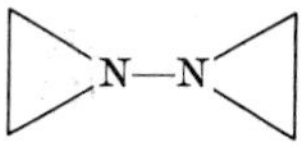	91
2	2	3; 5	Heteroatoms: nitrogen	
			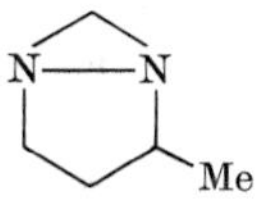	88

(*continued*)

43

TABLE IV (*continued*)

R	H	N	Compound	Reference[b]
2	2	5; 5	Heteroatoms: nitrogen	
			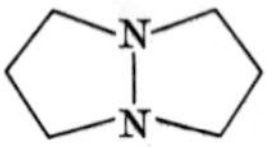	89, 91
			Heteroatoms: nitrogen, boron	
			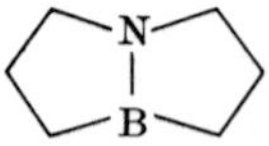	136, 140
2	2	5; 5	Heteroatoms: sulfur	
			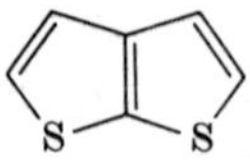	137, 156
			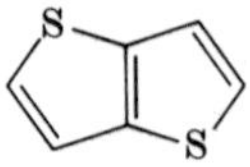	137, 156
2	2	5; 6	Heteroatoms: oxygen	
			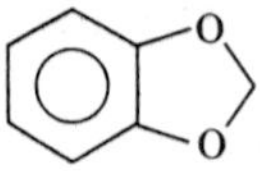	141
			Heteroatoms: nitrogen, sulfur	
			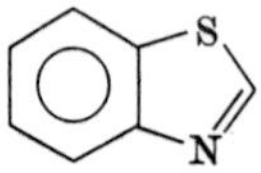	46
2	2	5; 7	Heteroatoms: nitrogen	
			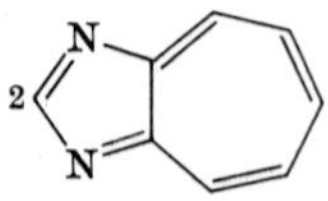	130
			2-NH_2	130
			2-$N(CH_3)_2$	130
			2-OCH_2CH_3	130
2	2	6; 6	Heteroatoms: nitrogen	
			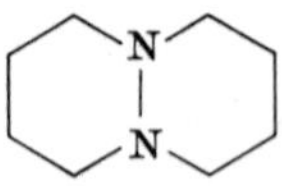	89

(continued)

TABLE IV (*continued*)

R	H	N	Compound	Reference[b]
2	2	6; 6	Heteroatoms: nitrogen	

			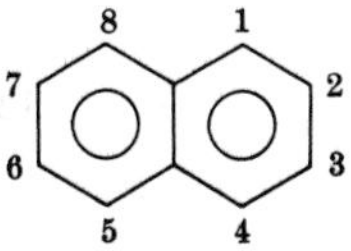 Compound	Reference[b]
			1,2-Diaza	15, 37, 51, 67
			1,3-Diaza	1, 37, 51, 67
			1,4-Diaza	15, 37, 51, 67
			1,5-Diaza	67
			1,6-Diaza	67
			1,7-Diaza	67
			1,8-Diaza	67
			2,3-Diaza	37, 67
			2,6-Diaza	67
			2,7-Diaza	67

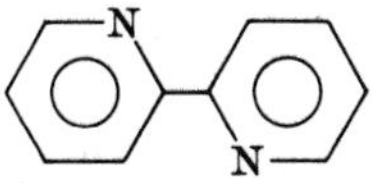

86

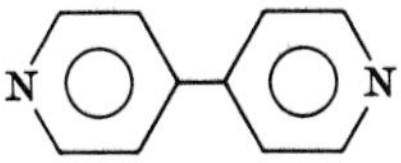

86

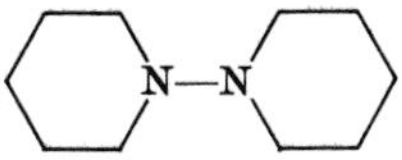

91

| 2 | 3 | 5; 5 | Heteroatoms: sulfur | |

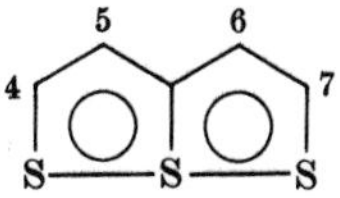

142

			4-Methyl	142
			4,7-Dimethyl	142
			4,7-Diphenyl	142
			5,6-Dimethyl	142
			5,6-Diphenyl	142

| 2 | 3 | 5; 6 | | |

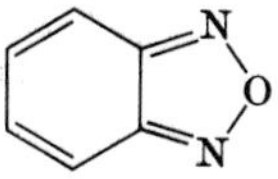

137

(*continued*)

TABLE IV (*continued*)

R	H	N	Compound	Reference[b]
2	3	5; 6	Heteroatoms: sulfur	
			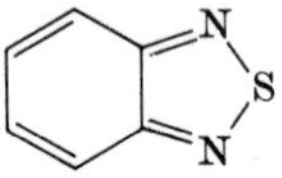	137, 156
			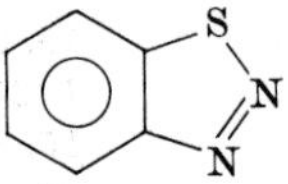	156
2	4	5; 5		143
			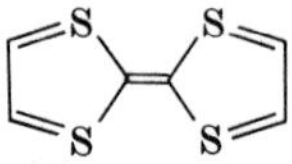	144
2	4	6; 6		136, 140
3	1	6; 5; 6		46
			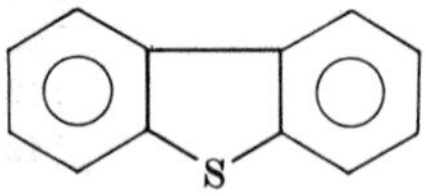	156
			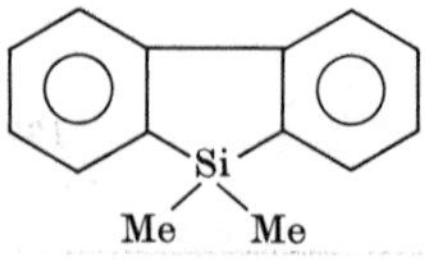	145

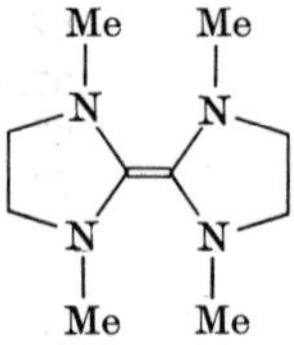

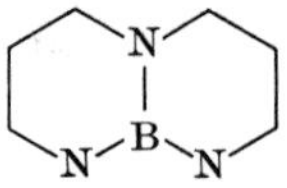

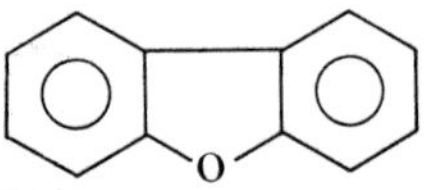

(*continued*)

TABLE IV (*continued*)

R	H	N	Compound	Reference[b]
3	1	6; 6; 6	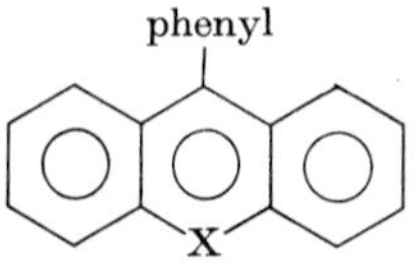	

phenyl

Heteroatom: nitrogen		72
phosphorus		72
arsenic		72

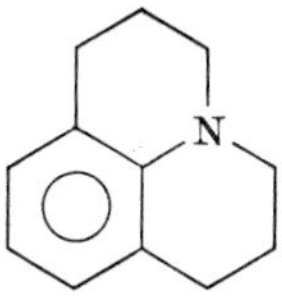

82

3	3	5; 5; 6	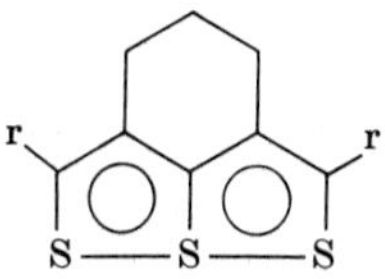	

r = Methyl	142
= Ethyl	142
= Isopropyl	142

4	1		

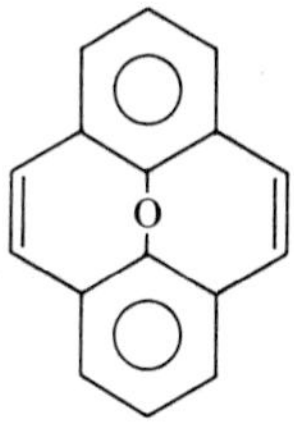

146

6	1 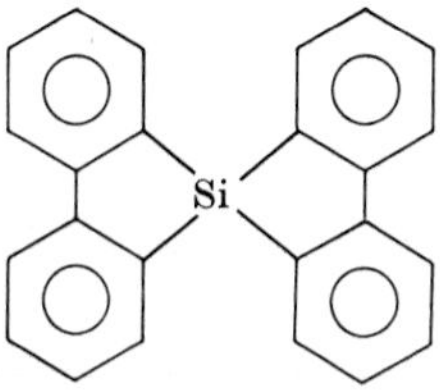		

145

(*continued*)

TABLE IV (*continued*)

H	Compound	Reference[b]
	Cage Molecules	

Bicyclo   1

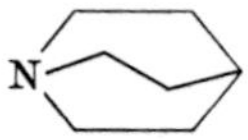

52

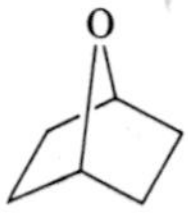

118

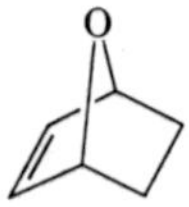

118

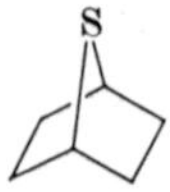

159

2

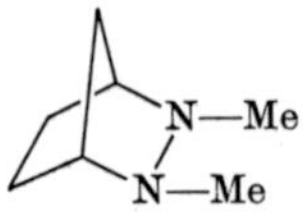

89

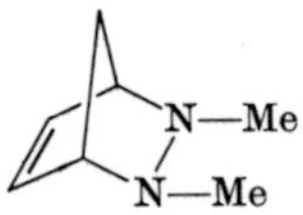

89

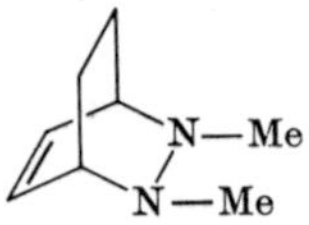

89

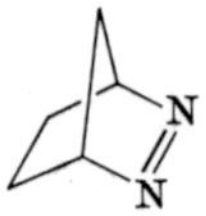

147, 160

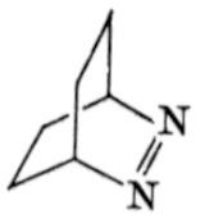

160

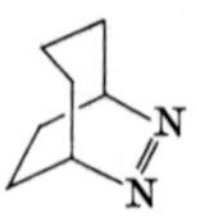

160

(*continued*)

TABLE IV (*continued*)

H	Compound	Reference[b]

| Bicyclo | 2 | | |

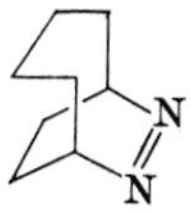

160

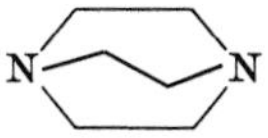

52, 53

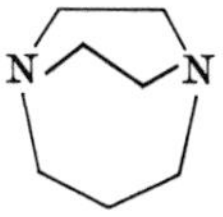

155

Tricyclo

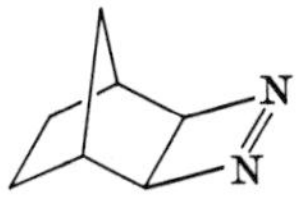

15, 147

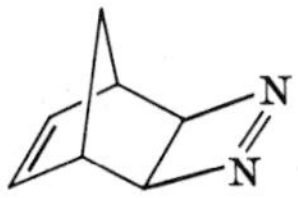

15, 147

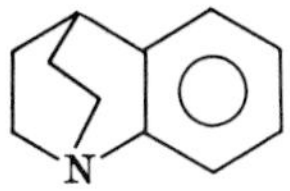

82

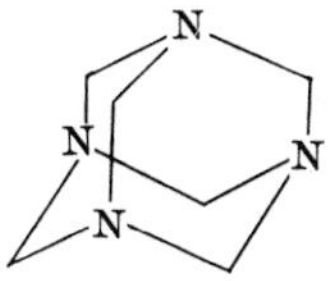

151

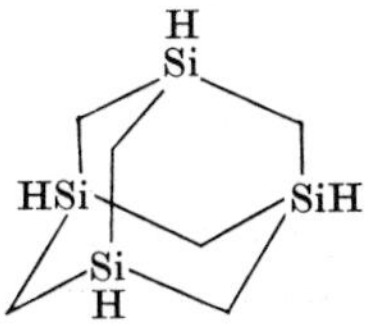

151

(*continued*)

TABLE IV (*continued*)

H	Compound	Reference[b]
	Miscellaneous	

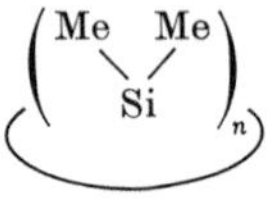

$$n = 5,\ 6 \qquad 148$$

	Reference
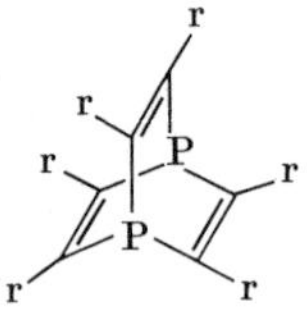	68
$r = $ methyl	68
$= CF_3$	68
$[NPMe_2]_n$	149
$n = 3,4$	
$[NP(NMe_2)_2]_n$	149
$n = 3,4$	
$[NPF_2]_n$	149
$n = 3,4,5,6,7,8$	
$[NPCl_2]_3$	149
$[NP(OMe)_2]_n$	149
$n = 3,4$	
$[NP(OCH_2CF_3)_2]_n$	149
$n = 3,4$	
$[NP(O\text{-phenyl})_2]_n$	149
$n = 3,4$	

[b] References:

[108] H. Basch, M. B. Robin, N. A. Kuebler, C. Baker, and D. W. Turner, *J. Chem. Phys.* **51**, 52 (1969).

[109] N. Bodor, M. J. S. Dewar, W. B. Jennings, and S. D. Worley, *Tetrahedron* **26**, 4109 (1970).

[110] A. D. Baker, D. Betteridge, N. R. Kemp, and R. E. Kirby, *Anal. Chem.* **43**, 375 (1971).

[111] D. C. Frost, F. G. Herring, A. Katrib, and C. A. McDowell, *Chem. Phys. Lett.* **20**, 404 (1973).

[112] H. Bock and B. Solouki, *Angew. Chem.* **84**, 436 (1972).

[113] A. D. Bain and D. C. Frost, *J. Electron. Spectrosc.* (in press).

[114] G. Hvistendahl and K. Undheim, *Chem. Scripta* **1**, 123 (1972).

[115] A. D. Baker, D. Betteridge, N. R. Kemp, and R. E. Kirby, *Chem. Commun.* 286 (1970).

[116] A. D. Baker, D. Betteridge, N. R. Kemp, and R. E. Kirby, *Anal. Chem.* **42**, 1064 (1970).

[117] M. Beez, Dissertation, Johann Wolfgang Göthe-Univ., Frankfurt am Main (1971).

with reference to Fig. 8 this is a necessary consequence of the through-bond interaction between the semilocalized n_+ orbital [c.f., (17)] and the CC–σ orbitals of appropriate symmetry.

In fact, observed ΔI and ΔE values can thus be used in view of an "experimental" assessment of relative orbital shapes.

ACKNOWLEDGMENTS

This work is part No. 56 of project No. 2.477.71 of the Schweizerischer Nationalfonds who has supported the research underlying the contributions from our institute. We gratefully acknowledge the generous help of Ciba-Geigy S.A., F. Hoffmann-La Roche Inc. S.A., and of Sandoz S.A. J. P. Maier thanks the Royal Society for a fellowship.

[118] A. D. Bain, J. C. Bünzli, D. C. Frost, and L. Weiler, *J. Amer. Chem. Soc.* **95**, 291 (1973).

[119] P. J. Derrick, L. Åsbrink, O. Edqvist, and E. Lindholm, in print.

[120] D. W. Turner, *in* "Molecular Spectroscopy" (P. Hepple, Ed.), p. 209. Inst. of Petroleum, London, 1968.

[121] J. W. Rabalais, L. O. Werme, T. Bergmark, L. Karlson, and K. Siegbahn, *Int. J. Mass Spectrom. Ion Phys.* **9**, 185 (1972).

[122] P. J. Derrick, L. Åsbrink, O. Edqvist, B.-O. Jonsson, and E. Lindholm (in press).

[123] A. Schweig, U. Weidner, and G. Manuel, *Angew. Chem.* **84**, 899 (1972).

[124] I. Morishima, K. Yoshikawa, T. Yonezawa, and H. Matsumoto, *Chem. Phys. Lett.* **16**, 336 (1972).

[125] J. N. Murrell and R. J. Suffolk, *J. Electron Spectrosc.* **1**, 471 (1972/73).

[126] J. Daintith, R. Dinsdale, J. P. Maier, D. A. Sweigart, and D. W. Turner, *in* "Molecular Spectroscopy" (P. Hepple, Ed.). Inst. of Petroleum, London, 1972.

[127] M. B. Robin, C. R. Brundle, N. A. Kuebler, G. B. Ellison, and K. B. Wiberg, *J. Chem. Phys.* **57**, 1758 (1972).

[128] E. Haselbach, E. Heilbronner, A. Mannschreck, and W. Seitz, *Angew. Chem.* **82**, 879 (1970).

[129] C. R. Brundle, M. B. Robin, N. A. Kuebler, and H. Basch, *J. Amer. Chem. Soc.* **94**, 1451 (1972).

[130] E. Heilbronner and T. Hoshi, *Z. Phys. Chem.* (*Neue Folge*), (in press).

[131] J. Kroner, W. Strack, and W. Kosbahn, *Z. Naturforsch.* **28b**, 188 (1973).

[132] D. R. Lloyd and N. Lynaugh, *Phil. Trans. Roy. Soc. London A* **268**, 97 (1970).

[133] D. C. Frost, F. G. Herring, C. A. McDowell and I. A. Stenhouse, *Chem. Phys. Lett.* **5**, 291 (1970).

[134] H. Bock and W. Fuss, *Angew. Chem.* **83**, 169 (1971).

[135] J. Kroner, D. Proch, W. Fuss, and H. Bock, *Tetrahedron* **28**, 1585 (1972).

[136] W. Fuss, Dissertation, Johann Wolfgang Goethe-Univ., Frankfurt am Main (1971).

[137] P. A. Clark, R. Gleiter, and E. Heilbronner, *Tetrahedron* **29**, 3085 (1973).

[138] D. M. V. Van den Ham and D. Van der Meer, *Chem. Phys. Lett.* **15**, 549 (1972).

[139] J. H. D. Eland and C. J. Danby, *Z. Naturforsch.* **23a**, 355 (1968).

[140] H. Bock and W. Fuss, *Chem. Ber.* **104**, 1687 (1971).

[141] P. C. Meier and W. Simon, *in* "Molecular Spectroscopy," p. 53. Inst. of Petroleum, London, 1971.

[142] R. Gleiter, V. Hornung, B. Lindberg, S. Högberg, and N. Lozac'h, *Chem. Phys. Lett.* **11**, 401 (1971).

143 B. Cetinkaya, G. H. King, S. S. Krishnamurthy, M. F. Lappert, and J. B. Pedley, *Chem. Commun.* 1370 (1971).

144 R. Gleiter, E. Schmidt, D. O. Cowan, and J. P. Ferraris, *J. Electron Spectrsoc.* **2**, 207 (1973).

145 A. Schweig, U. Weidner, D. Hellwinkel, and W. Krapp, *Angew. Chem.* **85**, 360 (1973); *Angew. Chem. Int. Ed.* **12**, 310 (1973).

146 R. Boschi and W. Schmidt, *Angew. Chem.* **83**, 403 (1973); *Angew .Chem. Int. Ed.* **12**, 402 (1973).

147 F. Brogli, W. Eberbach, E. Haselbach, E. Heilbronner, V. Hornung, and D. M. Lemal, *Helv. Chim. Acta* **56**, 1933 (1973).

148 H. Bock and W. Ensslin, *Angew. Chem.* **83**, 435 (1971).

149 G. R. Branton, C. E. Brion, D. C. Frost, K. A. R. Mitchell, and N. L. Paddock, *J. Chem. Soc. A* 151 (1970).

150 S. Cradock, R. H. Findlay, and M. H. Palmer, *Tetrahedron* **29**, 2173 (1973).

151 W. Schmidt, *Tetrahedron* **29**, 2129 (1973).

152 K. H. Ostoja Starzewski, H. tom Dieck and H. Bock, *J. Organometallic Chem.* (in press).

153 G. Wagner and H. Bock, *Chem. Ber.* **107**, 68 (1974).

154 W. Schäfer, A. Schweig, G. Märkl, H. Hauptmann, and F. Mathey, *Angew. Chem.* **85**, 140 (1973).

155 Y. Hamada, A. Y. Hirakawa, M. Tsuboi, and H. Ogata, *Bull. Chem. Soc. Jap.* **46**, 2244 (1973).

156 R. A. W. Johnstone and F. A. Mellon, *J. Chem. Soc. Faraday II* 1155, (1973).

157 W. Schäfer, A. Schweig, G. Märkl and K.-H. Heier, *Tetrahedron Lett.* 3743 (1973).

158 T. Kobayashi and S. Nagakura, *Bull. Chem. Soc. Jap.* **46**, 1558 (1973).

159 J. C. Bünzli, D. C. Frost, and L. Weiler, *J. Amer. Chem. Soc.* **95**, 7880 (1973).

160 R. J. Boyd, J. C. Bünzli, J. P. Snyder, and M. L. Heyman, *J. Amer. Chem. Soc.* **95**, 6478 (1973).

161 P. D. Mollere, *Tetrahedron Lett.* 2791 (1973).

162 W. Schäfer, A. Schweig, S. Gronowitz, A. Taticchi, and F. Fringuelli, *Chem. Commun.* 541 (1973).

163 G. Distefano, S. Pignataro, G. Innorta, F. Fringuelli, G. Marino, and A. Taticchi, *Chem. Phys. Lett.* **22**, 132 (1973).

164 A. Schweig and W. Thiel, *Chem. Phys. Lett.* **21**, 541 (1973).

165 A. D. Bain and D. C. Frost, *Can. J. Chem.* **51**, 1245 (1973).

· 2 ·

Microwave Spectroscopy of Heterocyclic Molecules

J. SHERIDAN

UNIVERSITY COLLEGE OF NORTH WALES
BANGOR, U.K.

I. INTRODUCTION

A. The Chief Features of Microwave Spectroscopy

The rotational spectroscopy of gaseous dipolar molecules is described in standard works.[1-4] Heterocyclic molecules are sufficiently heavy to be comprehensively studied in the popular frequency range from 8 to 40 GHz. Because molecular collisions broaden the absorptions, sample pressures are only some 10^{-2}–10^{-3} torr. Since it is not difficult to raise the temperature of the absorption cells well above that of the laboratory when necessary, a very wide range of substances can be studied. A material as nonvolatile as imidazole, for example, gives readily observed microwave spectra at room temperature.

Because a permanent molecular dipole moment is necessary for absorption, a few heterocyclic substances which have central symmetry, such as pyrazine or s-triazine, do not exhibit microwave spectra, but most such substances have strong spectra on account of the polarity induced by their heteroatoms. Virtually all heterocyclic molecules are asymmetric rotors, and each thus has its characteristic panorama of spectral lines displayed with huge resolution and highly accurate frequency measurement. Microwave spectra have become a rich source of knowledge of minute details of molecular structures (Section B), but contain no features analogous to group frequencies; they yield only information about the molecule as a whole, although this may tell us more about certain regions of the molecule than about others.

B. Types of Molecular Information Obtained

1. Molecular Geometry

The spectra are primarily analyzed in terms of accurate average values of the reciprocals of the three molecular moments of inertia, I_A, I_B, and I_C, where the subscripts refer to the principal axes of inertia of the molecule.[1-4] This generally leads at once to the characterization of the general molecular conformation, and some precise structural features may emerge. For example, if one of the moments of inertia is

[1] C. H. Townes and A. L. Schawlow, "Microwave Spectroscopy." McGraw-Hill, New York, 1955.

[2] T. M. Sugden and C. N. Kenney, "Microwave Spectroscopy of Gases." Van Nostrand Reinhold, Princeton, New Jersey, 1965.

[3] J. E. Wollrab, "Rotational Spectra and Molecular Structure." Academic Press, New York, 1967.

[4] W. Gordy and R. L. Cook, in "Technique of Organic Chemistry" (A. Weissberger, ed.), 2nd ed., Vol. 9. Wiley (Interscience), New York, 1970.

nearly exactly the sum of the other two ($I_C = I_A + I_B$), we recognize the property of a planar mass distribution, such as is found in many heterocyclic molecules, but exact interpretations of even such relationships must be made with caution, on account of the presence of molecular vibrations.

To derive individual structural parameters it is necessary to measure independent moments of inertia at least equal in number to these parameters. Since, in the cases concerned, there are more than three parameters, use has to be made of the different moments of inertia found for different isotopic forms of the molecule. For the molecules concerned here, the so-called substitution, or r_s-, geometries are the most accurate that can at present be obtained, and they are likely to remain so for some time. To obtain a complete structure in such ways it is necessary to measure the changes in moments of inertia which accompany the isotopic replacement of each atom in turn, when the location of each atom in the coordinate system of the A, B, and C axes of the "parent" isotopic form can be derived from relationships known as Kraitchman's equations. Except for any atoms which lie within about 0.2 Å of one of the axes, such isotopic changes of moments of inertia can be found with sufficient accuracy to determine atomic coordinates within a few thousandths of an angstrom, and internuclear distances with about the same accuracy; bond angles, by the substitution method, are commonly determined to within 0.2° or less. The isotopic substitution method is limited in its applicability by the fact that certain elements do not contain more than one stable isotope, but this is not the case for nearly all the atoms of interest in heterocyclic chemistry. The high sensitivity of microwave methods often allows carbon atoms to be located through spectra of species containing ^{13}C in its 1% natural concentration. Other atoms, such as hydrogen, and usually oxygen and nitrogen, require enrichments of ^{2}H, ^{18}O, and ^{15}N. Atoms close to principal axes may have certain coordinates at best poorly determined from isotopic shifts, although in certain ring structures it is possible to obtain large rotations of the axis system by isotopic substitution which allow selection of a "parent" species in which the axes are more favorably placed with respect to the atom of interest.

Atoms which are not directly located by substitution can be positioned, if there is sufficient information on the rest of the molecule, by placing them to allow the best overall fit on the measured moments of inertia or to allow the center of gravity of the molecule to be correctly located. Atoms so located will carry a somewhat larger error in position.

When the above methods do not locate all the atoms, certain

parameters may be assumed from knowledge of related molecules and bond radii. While many such assumptions can be made with reasonable confidence, the value of the data under such conditions will be reduced, and the effects of variations in the assumed parameters should be explored. Nevertheless, substitution methods give accurate locations of the substituted nuclei even if parts of the total structure may remain less well determined.

2. Electron Distribution

a. Magnitudes and Lines of Action of Electric Dipole Moments in the Molecular Framework. These are determined from accurate measurements of the Stark effect of assigned transitions, when electrostatic fields are applied to the gas. Analysis gives squared components of the dipole in the A, B, and C axes. For molecules with no axis of symmetry the sign ambiguity when the square roots of these quantities are taken leaves at this stage two or more possible directions of the total moment, but this direction may be derived unequivocally by redetermination of the squared dipole components for an isotopically substituted form of the molecule in which the inertial axes, but not of course the electron distribution, have been swung in direction. The line of action of the total moment may alternatively sometimes be assigned, with caution, from general chemical intuition. The sign of the moment, i.e., the orientation of the positive and negative charge separations in the molecular frame, can occasionally be determined experimentally from Zeeman effect measurements (Section I,B,2,c).

b. Nuclear Quadrupole Coupling Effects. The splittings of lines for molecules containing nuclei of spin quantum number greater than $\frac{1}{2}$ are analyzed in terms of the nuclear quadrupole coupling constants referred to the principal axes A, B, and C, designated eQq_{aa}, eQq_{bb}, and eQq_{cc}, respectively.[1-4] These are the products of the electronic charge e, the nuclear quadrupole moment Q, and the electric field divergences q_{aa}, q_{bb}, or q_{cc} at the nuclear location. The constants thus serve to indicate and compare different electronic environments of a particular nucleus in different molecules. The principal values and axes of the nuclear quadrupole coupling tensor within the molecular frame are often evident from the molecular symmetry, and in other cases they can be found from the effects of isotopic substitution on the coupling constants as the rotational axes are swung or, in the case of large coupling constants, from the measurement of second-order quadrupole coupling effects. Quadrupole splittings in spectra can be analyzed for

56

the case, commonly met in heterocyclic substances, where two or more quadrupole nuclei are present in the molecule.

The most important atom in heterocyclic structures displaying quadrupole coupling effects is ^{14}N. The coupling constants for this atom are unfortunately small, usually less than 5 MHz, and are not always measured with high accuracy, but they have nonetheless provided much information. Other quadrupole nuclei, particularly ^{35}Cl, ^{37}Cl, ^{79}Br, and ^{81}Br, are encountered as side group substituents.

The interpretation of quadrupole coupling data may still often usefully be made in terms of valence bond language and the p electron distribution as shown by Townes and Dailey, Gordy, and others.[1-4] More recently, coupling constants and complete tensors have been used as sensitive tests of molecular orbital descriptions in heterocyclic chemistry. The nuclear coupling tensor is only a critical test of the validity of an electron distribution close to the nucleus concerned, since the effects vary as the inverse cube of the electron–nuclear separation.

c. Information Derived from Zeeman Effects. The effects on rotation spectra of applying high magnetic fields to molecules have led to some of the most important recent advances in experimental knowledge of electron distributions. This development, due largely to W. H. Flygare and co-workers, has special significance for cyclic substances. The energy terms forming the frequency splittings contain chiefly a part proportional to the magnetic field (first-order Zeeman effect term) and a further contribution, normally much smaller, proportional to the square of the magnetic field (second-order Zeeman effect term). The powers of the method depend on the separate evaluation of the coefficients of both first and second-order terms, and this is accomplished by placing in a strong uniform magnetic field the entire absorption cell of a sensitive spectrometer with high resolving power. In particular, the second-order effects yield the anisotropies of the molecular susceptibility (χ), i.e., terms of the type $\chi_{cc} - \frac{1}{2}(\chi_{aa} + \chi_{bb})$ (subsequently called $\Delta\chi$), where the subscripts refer to the principal inertial axes. Such anisotropies for cyclic molecules are immediately significant and have an important bearing on any aromatic properties. Other parameters of the electron distribution are derived by combining the results for first- and second-order Zeeman effects with introduction of knowledge of the molecular geometry and, where appropriate, of the bulk magnetic susceptibility of the substance. Important data so obtained include the diagonal elements of the molecular quadrupole moment tensor and the average squared coordinates of electrons along particular axes (the so-called second moments of the electron distribution). From first-order

Zeeman effects in different isotopic forms of molecules, the signs of electric dipole moments may be derived (Section I,B,2,a), but most heterocyclic substances are too heavy for this method to be applied at present with the necessary precision. Details of the derivation of these and other types of information from Zeeman studies, and summarized data for many examples, will be found in the review by Flygare and Benson.[5]

3. Molecular Force Field Information

Microwave studies have provided important information about molecular force fields, particularly with reference to low frequency vibrational modes in cyclic structures. Spectra of molecules in states appreciably populated under the cell conditions are normally resolvable from the spectra of the ground state. The energy of vibrational excitation can be estimated, and the effects of vibration on the moments of inertia determined. Such work serves as a check on the symmetries and frequencies of the lowest modes of some heterocyclic molecules, and in particular it has contributed new knowledge of the potential functions for puckering of a number of rings.[6]

Also important are the barriers to internal rotation of methyl side groups attached to heterocyclic systems. A number of such barriers have now been determined by well-established methods involving the measurement of line splittings caused by the tunneling effect of the methyl hydrogens through the rotation barrier.[3,4]

C. Classification of Results for Individual Molecules

In the presentation of data for molecules in order of increasing ring size, the choice between saturated and aromatic forms as fundamental structures for five- and six-membered rings has been made on the basis of the greater information available for the aromatic systems. For conciseness, the estimated accuracies of specific molecular parameters have been omitted, but discussion takes account of the fuller details in the original papers.

It should be emphasized that these microwave findings closely resemble and complement similar information on homocyclic molecules, and that some mention of the findings for homocyclic structures is necessary in discussion. The definition of a heterocyclic substance is also somewhat arbitrary, and no discussion is given of the microwave work on cyclic structures formed by hydrogen bonding, by hydrogen

[5] W. H. Flygare and R. C. Benson, Mol. Phys. 20, 225 (1971).
[6] V. W. Laurie, Accounts Chem. Res. 3, 331 (1970).

bridges, or of the type met in π-cyclopentadienyl derivatives of metals. Important recent studies of carboranes containing carbon–boron rings[7-9] are also insufficiently related to the main field to be discussed.

II. THREE-MEMBERED RINGS

A. Molecular Geometry

Table I shows findings for ethylene oxide (oxirane), ethyleneimine (aziridine), ethylene sulfide, phosphirane, 1-chloroaziridine, ethylene episulfoxide, and ethylene episulfone. The dimensions of cyclopropane, obtained by electron diffraction, are included for comparison. For several of these substances, the structures are based on as complete isotopic substitution as is possible, but in others measurements are still to be made for certain isotopic substitutions, and this has been indicated by quoting the parameters to fewer significant figures. Such further measurements are desirable for complete comparisons.

Except for the unusually long C—C bond reported in ethylene episulfone, the C—C bonds in the heterocyclic rings are shorter than those in cyclopropane. The C—X bonds, again with the exception of those in ethylene episulfone, are generally longer than the corresponding bonds in molecules where the same hetero groups are joined to two methyl groups. The CXC angles, $2 \sin^{-1} (C—C/2C—X)$, are within 2° of 60° when X is O or N, and between 47° and 55° when X is S or P. The C—H lengths and values of $\angle HCH$ are similar to those in cyclopropane in the well-measured cases, phosphirane showing the longest CH and smallest $\angle HCH$. In ethyleneimine, 1-chloroaziridine, and phosphirane, in which the ring plane is no longer a plane of symmetry, the CH_2 planes are slightly tilted away from the exocyclic atom attached at the hetero group, but only in 1-chloroaziridine is any difference in CH lengths to the *syn*- and *anti*-positions apparent. In ethylene oxide, the CH_2 planes are tilted away from the bisector of $\angle CCO$, toward the oxygen, the C_2H_4 structure thus showing some of the flattening which is complete in ethylene. This tilt is measured by the amount by which $\angle H_2CC + \frac{1}{2} \angle CCO$ exceeds 180° ($\angle H_2CC$ is the angle between the CH_2 plane and the C—C bond). In ethylene oxide, this is nearly 8°. A comparable tilt in the same direction is reported in 1-chloroaziridine, and there is evidence of a similar but reduced effect in ethyleneimine, phosphirane and ethylene sulfide.

[7] R. A. Beaudet and R. L. Poynter, *J. Chem. Phys.* **53**, 1899 (1970).
[8] C. S. Cheung and R. A. Beaudet, *Inorg. Chem.* **10**, 1144 (1971).
[9] G. L. McKown and R. A. Beaudet, *Inorg. Chem.* **10**, 1350 (1971).

TABLE I

STRUCTURE PARAMETERS AND DIPOLE MOMENTS FOR THREE-MEMBERED RINGS[a]

Molecule	Distances (Å)				Angles in degrees			μ (D)	References
	CC	CX	XY	CH	∠HCH	∠H₂CC	∠YXC		
 ◁CH₂	1.510	1.510	1.089	1.089	115	150.0	117	0	b
 ◁O	1.466	1.431	—	1.085	116.6	158.0	—	1.88	c,d
 ◁NH	1.481	1.475	1.016	1.084	115.7	154	109.5	1.89	e,f
 ◁NCl	1.484	1.489	1.738	1.092 (syn) 1.079 (anti)	117.1	158	111.7	—	g
 ◁PH	1.502	1.867	1.428	1.093	114.4	149.5	95.2	1.12	h
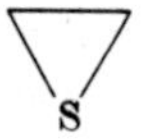 ◁S	1.484	1.815	—	1.083	115.8	151.8	—	1.84	c,l
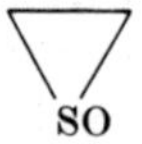 ◁SO	1.50	1.82	1.48	(a)	(a)	(a)	110	3.72	i
 ◁SO₂	1.59	1.73	1.44	(a)	(a)	(a)	117	4.45	j,k

[a] Hetero groups represented as X, XY, or XY₂; ∠H₂CC is the angle between the CC and CH₂ plane; (a) indicates that this parameter is assumed to be the same as in ethylene sulfide.

[b] O. Bastiansen, F. N. Fritsch, and K. Hedberg, *Acta Crystallogr.* **17**, 538 (1964).

[c] G. L. Cunningham, A. W. Boyd, R. J. Myers, W. D. Gwinn, and W. I. Le Van, *J. Chem. Phys.* **19**, 676 (1951).

[d] C. Hirose, *Bull Chem. Soc. Japan* (1974) (in press).

[e] B. Bak and S. Skaarup, *J. Mol. Struct.* **10**, 385 (1971).

[f] R. D. Johnson, R. J. Myers, and W. D. Gwinn, *J. Chem. Phys.* **21**, 1425 (1953).

[g] B. Bak and S. Skaarup, *J. Mol. Struct.* **12**, 259 (1972).

[h] M. T. Bowers, R. A. Beaudet, H. Goldwhite, and R. Tang, *J. Amer. Chem. Soc.* **91**, 17 (1969).

[i] S. Saito, *Bull. Chem. Soc. Japan* **42**, 663 (1969).

[j] Y. Nakano, S. Saito, and Y. Morino, *Bull. Chem. Soc. Japan* **43**, 368 (1970).

[k] H. Kim, *J. Chem. Phys.* **57**, 1075 (1972).

[l] K. Okiye, C. Hirose, D. G. Lister, and J. Sheridan, *Chem. Phys. Lett.* **24**, 111 (1974).

Diazirine, $\overline{CH_2—N=N}$, has been similarly investigated.[10] The substitution N—N distance (1.228 Å) is close to that for a double bond, and the substitution N—C distance (1.482 Å) is a little longer than the C—N bond in dimethylamine. The less precise CH_2 dimensions are similar to those in ethylene oxide.

Few internuclear distances have been determined by the microwave method in the substituted three-membered rings mentioned in Sections II,B and II,C. The data for these accord with dimensions predictable from the fundamental ring structures, this agreement extending to two or three isotopic forms in a few cases.

B. Electron Distribution

Dipole moments are also listed in Table I. Where the hetero group is asymmetric with respect to the ring plane, the moment makes an angle with this plane. This angle is about 30° for ethyleneimine and phosphirane and most probably 39° for ethylene episulfoxide, although an angle of 11° would also fit the data. The dipole moment of diazirine[10] is 1.59 D. Dipole moments have been measured for epoxybutyne[11] and for nearly all the substituted three-membered rings mentioned in this section and Section II,C; the moments indicate little interaction between substituents and the ring structures.

A number of nuclear quadrupole coupling constants have been measured. The ^{14}N coupling tensor in ethyleneimine[12,13] and that for [33]S in ethylene sulfide[14] are shown to be in reasonable general agreement with orbital descriptions. The ^{14}N coupling in diazirine[15] accords with the expected structure, as do the ^{14}N couplings in the substituted diazirines mentioned in Section II,C. The quadrupole couplings of chlorine and bromine in methylchloro-[16] and methylbromodiazirine[17] accord with only some 2–3% double bond character in the carbon–halogen bonds.

Detailed Zeeman effect studies have been made of ethylene oxide,[18,19]

[10] L. Pierce and V. Dobyns, *J. Amer. Chem. Soc.* **84**, 2651 (1962).
[11] M. J. Collins and J. E. Boggs, *J. Chem. Phys.* **57**, 3811 (1972).
[12] W. M. Tolles and W. D. Gwinn, *J. Chem. Phys.* **42**, 2253 (1965).
[13] M. K. Kemp and W. H. Flygare, *J. Amer. Chem. Soc.* **90**, 6267 (1968).
[14] R. L. Shoemaker and W. H. Flygare, *J. Amer. Chem. Soc.* **90**, 6263 (1968).
[15] J. M. Pochan and W. H. Flygare, *J. Phys. Chem.* **76**, 2249 (1972).
[16] J. E. Wollrab and L. H. Scharpen, *J. Chem. Phys.* **51**, 1584 (1969).
[17] J. E. Wollrab, *J. Chem. Phys.* **53**, 1543 (1970).
[18] D. H. Sutter, W. Hüttner, and W. H. Flygare, *J. Chem. Phys.* **50**, 2869 (1969).
[19] R. C. Benson and W. H. Flygare, *J. Chem. Phys.* **52**, 5291 (1970).

ethylene sulfide[19,20] and ethyleneimine.[21] Among the varied information derived, the magnetic susceptibility anisotropies, $\Delta\chi$, given by
$\chi_{cc} - \frac{1}{2}(\chi_{aa} + \chi_{bb})$, are the most interesting. In all molecules in this
review, the C-axis is perpendicular, or nearly perpendicular, to the
ring plane. A negative value of $\Delta\chi$ accordingly indicates a greater
diamagnetism with respect to a field impressed perpendicular to the
ring plane than the average diamagnetism with respect to fields impressed at mutually perpendicular directions in the ring plane. It is of
great interest that the magnetic anisotropies of many substances can
be reasonably expressed as the sums of local group anisotropies.[22,23]
These summation rules do not apply in the case of many cyclic structures. For ethylene oxide, ethylene sulfide, and ethyleneimine the
group anisotropies predict $\Delta\chi$ values close to zero. The observed values
are respectively (in 10^{-6} erg G^{-2} mole^{-1}) -9.4, -15.4, and -10.6.
These, and other three-membered rings, thus appear more diamagnetic
to fields impressed perpendicular to the ring plane than expected on the
basis of local group anisotropies. In larger rings this effect can be used
to estimate aromatic character, but this is not relevant to the description of nonlocal effects in the substances discussed here. The increased
negative values of $\Delta\chi$ must arise from an increased in-plane electron
density or the absence of low-lying unoccupied electronic states such
as lead to the opposite effects in four-membered rings (Section III,B).[24]
The above effects are seen particularly clearly when the $\Delta\chi$ values are
compared with those for $(CH_3)_2O$ and $(CH_3)_2S$ with the C-axes still
perpendicular to the COC and CSC planes; these open chain substances
have $\Delta\chi$ values[19] (in 10^{-6} erg G^{-2} mole^{-1}) of $+4.6$ and $+3.5$,
respectively.

Much further discussion of the electronic structures of three-
membered heterocyclic substances will be found in the papers cited. In
the simplest cases it is reasonable to expect SCF MO calculations, already appearing,[25] to form ultimately the best basis of interpretation.

C. Molecular Force Fields

Inversion of bonds about N or P in ethyleneimine and phosphirane
is known, from failure to resolve inversion splittings, to be opposed by
relatively high barriers. In ethyleneimine[12,13] the barrier is at least

[20] D. H. Sutter and W. H. Flygare, *Mol. Phys.* **16**, 153 (1969).
[21] D. H. Sutter and W. H. Flygare, *J. Amer. Chem. Soc.* **91**, 6895 (1969).
[22] T. G. Schmalz, C. L. Norris, and W. H. Flygare, *J. Amer. Chem. Soc.* **95**, 7961 (1973).
[23] R. C. Benson and W. H. Flygare, *J. Chem. Phys.* **58**, 2651 (1973).
[24] C. L. Norris, H. L. Tigelaar, and W. H. Flygare, *Chem. Phys.* **1**, 1 (1973).
[25] R. Bonaccorsi, E. Scrocco, and J. Tomasi, *J. Chem. Phys.* **52**, 5270 (1970).

11.6 kcal mole^{-1}, and may be much higher; it is thus some two or more times the inversion barrier in ammonia, which fact accords with the smaller interbond angles in ethyleneimine and contrasts with the opposite situation found in pyrrole and related aromatic five-ring compounds (Section IV,A).

Barriers opposing internal rotation of methyl groups attached to three-membered rings have now been found from a number of microwave studies. The threefold barrier heights (V_3) are obtained with a precision of about $\pm 10\%$ and occasionally better, the uncertainty being mostly due to possible small errors in the geometrical parameters. Values (in cal mole^{-1}) are: propylene oxide,[26,27] 2560 (confirmed by relative intensity studies[28,29]); cis-2,3-epoxybutane,[30] 1607; trans-2,3-epoxybutane,[31] 2444; propylene sulfide,[32] 3240; trans-propyleneimine,[33-35] 2608 (infrared spectra[36] give 2970 for the cis form); N-methyl ethyleneimine,[37] 3462; cis-propylene phosphine,[38] 3010; trans-propylene phosphine,[38] 3050; methyldiazirine,[39] 774; dimethyldiazirine,[40] 1129; methyl chlorodiazirine,[16] 1689; methyl bromodiazirine,[17] ~1700. In derivatives of two-carbon rings, the barriers are in the region of V_3 for methyl cyclopropane,[41] 2860 cal mole^{-1}, although the methyl groups in cis-2,3-epoxybutane rotate somewhat more freely. The barrier in methyl diazirine is low, but much higher values are found when a second substituent is present on the ring carbon. The equilibrium conformation of the methyl group in methyl diazirine is

[26] J. D. Swalen and D. R. Herschbach, J. Chem. Phys. **27**, 100 (1957).

[27] D. R. Herschbach and J. D. Swalen, J. Chem. Phys. **29**, 761 (1958).

[28] A. S. Esbitt and E. B. Wilson, Rev. Sci. Instrum. **34**, 901 (1963).

[29] A. N. Aleksandrov and G. I. Tysovskii, Zh. Strukt. Khim. **8**, 76 (1967).

[30] M. L. Sage, J. Chem. Phys. **35**, 142 (1961).

[31] M. R. Emptage, J. Chem. Phys. **47**, 1293 (1967).

[32] S S. Butcher, J. Chem. Phys. **38**, 2310 (1963).

[33] D. Hayes and E. L. Beeson, J. Chem. Phys. **47**, 3717 (1967).

[34] Y. S. Li, M. D. Harmony, D. Hayes, and E. L. Beeson, J. Chem. Phys. **47**, 4514 (1967).

[35] C. F. Su and E. L. Beeson, J. Chem. Phys. **59**, 759 (1973); see also R. Schmidt and E. L. Beeson, J. Chem. Phys. **59**, 3070 (1973); Y. S. Li, and J. R. Durig, J. Mol. Spectrosc. **47**, 179 (1973).

[36] J. R. Durig, S. F. Bush, and W. C. Harris, J. Chem. Phys. **50**, 2851 (1969).

[37] M. D. Harmony and M. Sancho, J. Chem. Phys. **47**, 1911 (1967).

[38] M. T. Bowers, R. A. Beaudet, H. Goldwhite, and S. Chan, J. Chem. Phys. **52**, 2831 (1970).

[39] L. H. Scharpen, J. E. Wollrab, D. P. Ames, and J. A. Merritt, J. Chem. Phys. **50**, 2063 (1969).

[40] J. E. Wollrab, L. H. Scharpen, D. P. Ames, and J. A. Merritt, J. Chem. Phys. **49**, 2405 (1968).

[41] R. G. Ford and R. A. Beaudet, J. Chem. Phys. **48**, 4671 (1968).

shown[42] to be the expected one in which a CCH plane bisects the N=N bond. In the case of *cis*- and *trans*-propylene phosphines,[38] relative intensities of spectra show the *trans* form to be the stabler by about 130 cal mole^{-1}.

When asymmetric substituents such as CH_2F or CH_2Cl are attached to a ring such as ethylene oxide, three different rotational isomers are possible. Spectra of epifluorohydrin[43] indicate predominantly a form in which one HCC plane approximately bisects the ring and a C—O bond, the O and F atoms being in *gauche* relative positions. In epichlorohydrin[44] evidence of the presence of all three conformers has been reported. Similar studies with other asymmetric substituents are likely soon.

III. FOUR-MEMBERED RINGS

A. Molecular Geometry

Most of the interest in the geometry of these substances concerns the planarity or otherwise of the ring systems (Section III,C). In only a few cases have individual parameters been established by isotopic substitution, and then not to the extent of full structure determination. In trimethylene oxide[45] substitutions were made at H and O atoms and the carbon atoms were placed by less direct methods (see Section I,B,1) to give a C—C length of 1.549 Å and a C—O length of 1.449 Å. Both these lengths are somewhat longer than in propane or dimethyl ether, respectively, and C—C is the same as found in cyclobutane by electron diffraction;[46] some lengthening over the normal C—C length may reflect the repulsions between methylene groups. In trimethylene oxide C—H is near 1.09 Å (also found in cyclobutane) and ∠HCH is near 110.5°, considerably less than in ethylene oxide. The ring angles in trimethylene oxide are not equal; that at the central carbon is only 84.6° and the other three angles are near 91.8°. When the central methylene group in trimethylene oxide is replaced by C=O to give 3-oxetanone,[47] the substitution lengths of the C—C bonds are 1.522 Å and the ∠CCC is 88.1°.

[42] D. W. Gord and J. E. Wollrab, *J. Chem. Phys.* **51**, 5728 (1969).

[43] S. C. Dass, A. Bhaumik, W. V. F. Brooks, and R. M. Lees, *J. Mol. Spectrosc.* **38**, 281 (1971).

[44] F. Ito, J. C. Chang, and H. Kim, *Symp. Mol. Struct. Spectrosc., Columbus, Ohio* Paper B8 (1971).

[45] S. I. Chan, J. Zinn, and W. D. Gwinn, *J. Chem. Phys.* **34**, 1319 (1961).

[46] A. Almenningen, O. Bastiansen, and P. N. Skancke, *Acta. Chem. Scand.* **15**, 711 (1961).

[47] J. S. Gibson and D. O. Harris, *J. Chem. Phys.* **57**, 2318 (1972).

B. Electron Distribution

Table II includes the dipole moments, and their lines of action, in the cases studied. These data conform generally with expectation.

Zeeman effect studies have been made on trimethylene oxide,[48] trimethylene sulfide,[48] 1,3-propiolactone,[49] diketene,[49] 3-oxetanone,[24] and 3-methylene oxetane[24] and numerous parameters have been determined (Section I,B,2,c). The magnetic susceptibility anisotropies, $\Delta\chi$, are listed in Table II, with the values predicted from group anisotropies, $(\Sigma\Delta X_{gp})$.[22,24] In these and other four-membered rings[24] the $\Delta\chi$ values are considerably more positive than predicted from local group contributions, corresponding to an increased paramagnetism with respect to a field impressed perpendicular to the ring. This property, which distinguishes four-membered rings from all other rings so studied, is ascribed[24] to the existence, in four-membered rings, of unoccupied electronic levels not far above the ground state.

C. Molecular Force Fields

Nearly all the interest in microwave work on four-membered rings has centered on the planarity or otherwise of the rings and the potential functions for the vibrations known as ring puckering. It is well known that in saturated four-membered rings there can be a close balance between the ring strain, which favors ring planarity, and methylene group H–H repulsions, which favor ring puckering. In the simplest examples, such as cyclobutane,[50] the equilibrium configuration is moderately puckered; passage from one puckered form to the other involves an energy barrier, arising from the repulsions at the planar configuration, while increasing ring strain limits the puckering to moderate values. The potential function for the puckering vibration is then of double minimum form. Structure changes which increase the dominance of ring strain over repulsions will decrease the puckering at the equilibrium configuration and may lead to planar ring geometries.

Ring-puckering vibrations are of low frequency and large amplitude and rotational spectra of molecules in the several lowest levels of the puckering mode can usually be assigned. The relative intensities of such spectra, and occasionally spectral frequencies, can give energies of puckering excitation, and the rotational constants (reciprocals of I_A, I_B, and I_C; Section I,B,1) are obtained for each puckering state.

[48] R. C. Benson, H. L. Tigelaar, S. L. Rock, and W. H. Flygare, *J. Chem. Phys.* **52**, 5628 (1970).

[49] H. L. Tigelaar, T. D. Gierke, and W. H. Flygare, *J. Chem. Phys.* **56**, 1966 (1972).

[50] F. A. Miller and R. J. Capwell, *Spectrochim. Acta* **27A**, 947 (1971).

TABLE II   ELECTRON DISTRIBUTIONS AND FORCE FIELDS IN FOUR-MEMBERED RINGS

Molecule	μ (D)	$\theta^{\circ a}$	$\Delta\chi$ (10^{-6} G^{-2} mole^{-1})	$\Sigma\Delta\chi_{gp}$ (10^{-6} G^{-2} mole^{-1})	V_1 (cal mole^{-1})	$b/a \times 10^{2b}$ (Å^2)	References
 (cyclobutane)	0	—	—	—	1480	−5.47	c
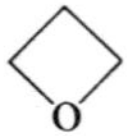 (O)	1.93	180	16.8	6.8	44	−0.91	d–g
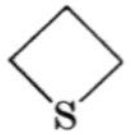 (S)	1.85	180	22.8	—	784	−4.66	g–j
 (Se)	—	—	—	—	1087	−5.97	k,l
 (SiH₂)	0.44	0			1260	−6.76	m,n
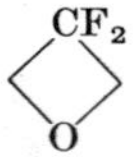 (CF₂)	0.40	(0) (180)			0	>0	o
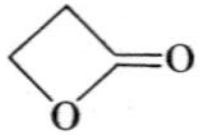 (lactone)	4.18	118	−0.6	−5.7	54	<0	p–s
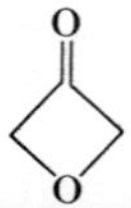 (dione)	0.89	0	−0.9	−5.7	0	+23.2	t,u

66

	1.63	180	4.3	−3.6	0	+10.2	u–w
	3.6	131	−7.2	−16.1	0	>0	s,x,y

[a] θ is the probable approximate angle between the direction of μ with the sign shown and the indicated near-symmetry axis of the four-membered ring: For high V_1 values, there is a dipole component perpendicular to the general ring plane.

[b] b/a in the expression $V_z = az^4 + bz^2$, where z is half the distance between the ring-diagonals, in Å.

[c] F. A. Miller and R. J. Capwell, Spectrochim. Acta 27A, 947 (1971).

[d] S. I. Chan, J. Zinn, J. Fernandez, and W. D. Gwinn, J. Chem. Phys. 33, 1643 (1960).

[e] S. I. Chan, T. R. Borgers, J. W. Russell, H. L. Strauss, and W. D. Gwinn, J. Chem. Phys. 44, 1103 (1966).

[f] R. A. Creswell and I. M. Mills, Eur. Microwave Spectrosc. Conf., 2nd, Bangor, Wales Paper 2A3 (1972).

[g] R. C. Benson, H. L. Tigelaar, S. L. Rock, and W. H. Flygare, J. Chem. Phys. 52, 5628 (1970).

[h] M. S. White and E. L. Beeson, J. Chem. Phys. 43, 1839 (1965).

[i] D. O. Harris, H. W. Harrington, A. C. Luntz, and W. D. Gwinn, J. Chem. Phys. 44, 3467 (1966).

[j] H. W. Harrington, J. Chem. Phys. 44, 3481 (1966).

[k] M. G. Petit, J. S. Gibson, and D. O. Harris, J. Chem. Phys. 53, 3408 (1970).

[l] A. B. Harvey, J. R. Durig, and A. C. Morrissey, J. Chem. Phys. 50, 4949 (1969).

[m] W. C. Pringle, J. Chem. Phys. 54, 4979 (1971).

[n] J. Laane and R. C. Lord, J. Chem. Phys. 48, 1508 (1968).

[o] G. L. McKown and R. A. Beaudet, J. Chem. Phys. 55, 3105 (1971).

[p] L. M. Boggia, P. G. Favero, and O. M. Sorarrain, Chem. Phys. Lett. 12, 382 (1971).

[q] N. Kwak, J. H. Goldstein, and J. W. Simmons, J. Chem. Phys. 25, 1203 (1956).

[r] D. W. Boone, C. O. Britt, and J. E. Boggs, J. Chem. Phys. 43, 1190 (1965).

[s] H. L. Tigelaar, T. D. Gierke, and W. H. Flygare, J. Chem. Phys. 56, 1966 (1972).

[t] J. S. Gibson and D. O. Harris, J. Chem. Phys. 57, 2318 (1972).

[u] C. L. Norris, H. L. Tigelaar, and W. H. Flygare, Chem. Phys. 1, 1 (1973).

[v] J. S. Gibson and D. O. Harris, J. Chem. Phys. 52, 5234 (1970).

[w] C. S. Blackwell and R. C. Lord, in "Vibrational Spectra and Structure" (J. R. Durig, ed.), Vol. 1, Chap. 1. Dekker, N. Y., 1972.

[x] F. Mönnig, H. Dreizler, and H. D. Rudolph, Z. Naturforsch. A 22, 1471 (1967).

[y] A. Bhaumik, W. V. F. Brooks, and S. C. Dass, Symp. Mol. Struct. Spectrosc., Columbus, Ohio Paper W8 (1970).

The spacing of the energy levels and the changes in rotational constants with vibrational quantum number lead to information on the puckering potential function; for a double minimum potential, the levels tend to approach in pairs and the rotation constants vary in a characteristic zig-zag way with vibrational quantum number. Usually the data are strongly supported by studies of these vibrations by infrared and Raman spectroscopy.[51-54]

The information obtained on puckering of rings is summarized in Table II by listing of the barriers to ring inversion (V_i) and some properties of the potential function. In nearly all cases the energy of the molecule, V_z, can be expressed as a function of z, a distance measuring the nonplanarity, usually half the perpendicular distance between the diagonals of the ring (or sometimes an angle of pucker or dimensionless reduced coordinate) in the following form: $V_z = az^4 + bz^2$. The energy, V_z, is zero at the planar configuration, and if the coefficient b is negative, the plot of V_z as a function of positive and negative values of z has a double minimum form. The ratios of the coefficients of quadratic and quartic terms, b/a, are listed in Table II. A higher barrier, V_i, is roughly associated with a higher negative value of b/a and a larger value of z at equilibrium. When b is positive, the puckering function has a single minimum at the planar configuration. In a very approximate sense, the az^4 term represents ring strain at increasing z values, and a negative bz^2 term the methylene H–H repulsions.

Cyclobutane[50] has a V_i of 1480 cal mole^{-1} and the ring diagonals are 0.33 Å apart at the equilibrium configuration. In trimethylene oxide the removal of two methylene hydrogens leads to a greatly reduced V_i; in this substance the ground vibrational level is some 35 cal mole^{-1} above the top of the barrier, and, in this sense, the average conformation of the ring is planar in all states. For trimethylene sulfide, trimethylene selenide, and silacyclobutane, V_i progressively increases in that order; bond angles at the hetero atoms in these substances may decrease the ring strain while retaining or even increasing the repulsion energies. These rings are puckered in their ground states.

The substituted trimethylene oxides listed in Table II have all been

[51] J. Laane, *Quart. Rev. (London)* **25**, 533 (1971).

[52] C. S. Blackwell and R. C. Lord, *in* "Vibrational Spectra and Structure" (J. R. Durig, ed.), Vol. 1, Chapter 1. Dekker, New York, 1972.

[53] J. Laane, *in* "Vibrational Spectra and Structure" (J. R. Durig, ed.), Vol. 1, Chapter 2. Dekker, New York, 1972.

[54] K. D. Möller and W. G. Rothschild, "Far-infrared Spectroscopy." Wiley (Interscience), New York, 1971.

reported to have single minimum puckering potential functions and planar ring geometry, but recent work on 1,3-butyrolactone supports the small V_1 listed. Since the double minimum property is weak in trimethylene oxide, it is not surprising to find no such property in substitution products, especially when the substituents, as mostly the case here, reduce the methylene H–H repulsion. A similar effect of such substituents in reducing V_1 is found in substituted cyclobutanes.[6,51]

IV. FIVE-MEMBERED RINGS WITH AROMATIC CHARACTER

Fundamental systems in which, in addition to one or more hetero atoms, there are formally two double bonds have been much studied. The planarity of these structures makes them particularly suited to detailed investigation of molecular geometry and electron distribution.

A. Molecular Geometry

Complete molecular structures by the substitution method have been found for some of the best known cases, largely through the work of the group at the University of Copenhagen. Some such structures are more readily determined than others. It is an advantage if the ring contains an axis of twofold symmetry. This reduces the number of parameters and allows readier detection of spectra due to low concentrations of species with isotopes located in positions which are equivalent with respect to the symmetry axis; the intensities of the spectra then correspond to twice the concentration of the species which would be present in the absence of molecular symmetry. Furthermore, in unsubstituted rings containing only atoms of similar mass (C, N, O) the A and B axes are swung through large angles by isotopic substitution, and it is normally possible to find isotopic forms in which these axes do not pass close to any atom. When a heavier atom (S, Se, Te) is present in an unsubstituted ring, the A axis in any isotopic form remains close to the direction from the ring center to the heaviest ring atom, but this is not a serious limitation in rings with twofold symmetry. When two or more hetero atoms are present in such a pattern that the ring no longer has a twofold symmetry axis, the detailed determination of structure becomes a longer task. When one atom such as S or Se is present in a ring of this type, the A axis must pass close to, but not exactly through, the heavy nucleus, and other atoms also lie near the B axis. Nonetheless, some structures of this type have now been fully determined, and work on others is well advanced.

J. SHERIDAN

TABLE III

STRUCTURE PARAMETERS IN AROMATIC FIVE-MEMBERED RINGS

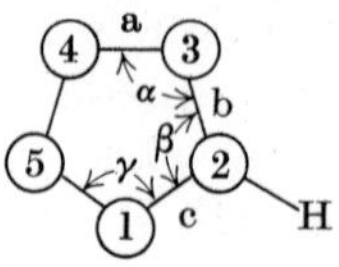

Molecule	Bond length (Å)			Angles in degrees				References
	a	b	c	α	β	γ	θ^a	
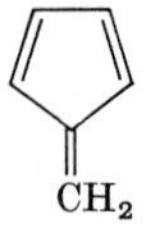 (cyclopentadienylidene, CH₂)	1.476	1.355	1.470	109.0	107.7	106.6	1.4	b
 (cyclopentadiene)	1.469	1.342	1.509	109.4	109.3	102.8	—	c
 (furan, O)	1.431	1.361	1.362	106.0	110.7	106.6	8.8	d,e
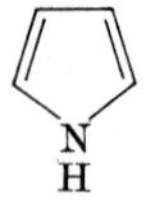 (pyrrole, N H)	1.417	1.382	1.370	107.4	107.7	109.8	4.6	f
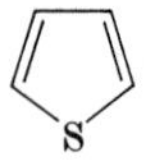 (thiophene, S)	1.423	1.370	1.714	112.4	111.5	92.2	4.5	g
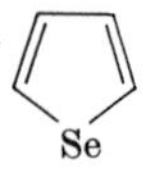 (selenophene, Se)	1.433	1.370	1.855	114.6	111.6	87.8	2.5	h,i
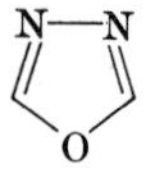 (N—N O)	1.399	1.297	1.348	105.6	113.4	102.0	5.2	j
 (N—N S)	1.371	1.302	1.721	112.2	114.6	86.4	0.8	k,l
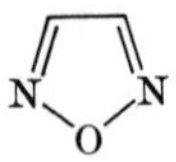 (N O N)	1.421	1.300	1.380	109.0	105.8	110.4	—	m

TABLE III (continued)

Molecule	Bond length (Å)			Angles in degrees				References
	a	b	c	α	β	γ	θ^a	
(structure)	1.420	1.328	1.631	113.8	106.4	99.6	—	n
(structure)	1.372	$\begin{cases}1.304 \\ (CN) \\ 1.367 \\ (CC)\end{cases}$	$\begin{cases}1.724 \\ \\ 1.713\end{cases}$	$\begin{cases}110.1 \\ \\ 115.8\end{cases}$	$\left.\begin{matrix}115.2 \\ \\ 109.6\end{matrix}\right\}$ 89.3	$\begin{matrix}1.1 \\ \\ 3.8\end{matrix}$		o
(structure)	1.416	$\begin{cases}1.331 \\ (CN) \\ 1.373 \\ (CC)\end{cases}$	$\begin{cases}1.349 \\ \\ 1.359\end{cases}$	$\begin{cases}111.9 \\ \\ 104.5\end{cases}$	$\left.\begin{matrix}104.1 \\ \\ 106.4\end{matrix}\right\}$ 113.1	$\begin{matrix}— \\ \\ 5.3\end{matrix}$		p–r

$^a\ \theta = 180° - (\tfrac{1}{2}\beta + \angle 12\text{H})$ or equivalent.

^{b} P. A. Baron, R. D. Brown, F. R. Burden, P. J. Domaille, and J. E. Kent, *J. Mol. Spectrosc.* **43**, 401 (1972).

^{c} L. H. Scharpen and V. W. Laurie, *J. Chem. Phys.* **43**, 2765 (1965).

^{d} B. Bak, D. Christensen, W. B. Dixon, L. Nygaard, J. R. Andersen, and M. Schottländer, *J. Mol. Spectrosc.* **9**, 124 (1962).

^{e} G. O. Sørensen, *J. Mol. Spectrosc.* **22**, 325 (1967).

^{f} L. Nygaard, J. T. Nielsen, J. Kirchheiner, G. Maltesen, J. R. Andersen, and G. O. Sørensen, *J. Mol. Struct.* **3**, 491 (1969).

^{g} B. Bak, D. Christensen, L. Nygaard, and J. R. Andersen, *J. Mol. Spectrosc.* **7**, 58 (1961).

^{h} N. M. Pozdeev, O. B. Akulinin, A. A. Shapkin, and N. N. Magdesieva, *Zh. Strukt. Khim.* **11**, 869 (1970).

^{i} R. D. Brown, F. R. Burden, and P. D. Godfrey, *J. Mol. Spectrosc.* **25**, 415 (1968).

^{j} L. Nygaard, R. L. Hansen, J. T. Nielsen, J. R. Andersen, G. O. Sørensen, and P. A. Steiner, *J. Mol. Struct.* **12**, 59 (1972).

^{k} B. Bak, L. Nygaard, E. J. Pedersen, and J. R. Andersen, *J. Mol. Spectrosc.* **19**, 283 (1966).

^{l} L. Nygaard, R. L. Hansen, and G. O. Sørensen, *J. Mol. Struct.* **9**, 163 (1971).

^{m} E. Saegebarth and A. P. Cox, *J. Chem. Phys.* **43**, 166 (1965).

^{n} V. Dobyns and L. Pierce, *J. Amer. Chem. Soc.* **85**, 3553 (1963).

^{o} L. Nygaard, E. Asmussen, J. H. Høg, R. C. Maheshwari, C. H. Nielsen, I. B. Petersen, J. R. Andersen, and G. O. Sørensen, *J. Mol. Struct.* **8**, 225 (1971); **9**, 220 (1971).

^{p} L. Nygaard, D. Christen, J. T. Nielsen, E. J. Pedersen, O. Snerling, E. Vestergaad, and G. O. Sørensen, *J. Mol. Struct.* (in press, 1974).

^{q} N. M. Pozdeev, R. S. Nasibullin, R. G. Latypova, V. G. Vinokurov, and N. D. Konevskaya, *Opt. Spektrosk.* **31**, 63 (1971).

^{r} R. S. Nasibullin, R. G. Latypova, V. S. Troitskaya, V. G. Vinokurov, and N. M. Pozdeev, *Opt. Spektrosk.* **35**, 92 (1973).

Table III summarizes the main findings for those examples which have been studied in detail. More discussion will be found in the original papers, and in a review by Nygaard.[55] We can simplify the presentation by remarking that the C—H distances are always within 0.005 Å of 1.075 Å, which is itself close to the C—H length in ethylene. Furthermore, except when there are hetero atoms at adjacent ring positions, C—H bonds at the 3- and 4-positions lie within one degree of the exterior bisector of the ring angles; if the C—H line deviates from this bisector at all, it would seem to do so very slightly in the sense that $\angle HC(3)C(4) > \angle HC(3)C(2)$. When hetero atoms are at adjacent ring positions, these hydrogens are displaced from the bisectors of the ring angles toward the adjacent hetero atom, and these cases will be mentioned individually. The C—H bonds at positions 2 and 5 are displaced from the bisectors of the ring angles toward atom 1 by angles which are shown as θ in Table III.

Most of the ring distances in Table III have been derived from isotopic substitutions at all positions, but in a few cases certain atoms have been placed from a knowledge of the positions of all other atoms and of the molecular center of gravity. Parameters dependent on the location of these atoms are subject to somewhat larger errors, but in general the distances may be taken as accurate to within a few thousandths of an angstrom and the angles to within about 0.2°. Also included in the table are the microwave findings for cyclopentadiene and fulvene.

Discussion is best based chiefly on the ring distances, since the ring angles may show effects arising simply from the need to close the ring. In cyclopentadiene and fulvene, the C(2)C(3)C(4)C(5) structure shows bond lengths quite close to those in 1,3-butadiene.[56] In the systems with only one hetero atom, the bond lengths in the corresponding four-carbon grouping show the expected shortening, in comparison with the homocyclic dienes, of the C(3)—C(4) bond, and corresponding lengthening of the C(2)—C(3) and C(4)—C(5) bonds; judged on this criterion, the amount of delocalization in this part of the molecule increases along the sequence furan, selenophene, thiophene, pyrrole. The related shortenings in the bonds joining the hetero atom to the C(2) and C(5), relative to the corresponding bond lengths in dimethyl derivatives of the hetero group, also increase in approximately this

[55] L. Nygaard, *J. Mol. Struct.* (to be published).

[56] K. Kuchitsu, *in* "Molecular Structure and Properties" *M.T.P. Int. Rev. Sci.* (G. Allen, ed.), Vol. 2, p. 203. Butterworth, London and Washington, D.C., 1972.

order. Delocalization in pyrrole, and also in pyrazole, imidazole,[57] 1,2,3-triazole,[58] and 1,2,4-triazole,[59] is sufficient to bring the imino hydrogen into the plane of the other atoms ($I_C - I_A - I_B \simeq 0.02$ amu Å^2 only in these molecules). The ring angle is close to 90° at sulfur and selenium, and the ring angles at the carbon atoms are correspondingly greater by a few degrees in thiophene and selenophene than in furan and pyrrole. In a similar way, the bond lengths suggest larger delocalization in 1,3,4-thiadiazole than in 1,3,4-oxadiazole. The ring angle at a pyridine-like nitrogen is generally lower than in the corresponding compound with a CH group in that position. The less complete data for 1,2,5-oxadiazole and 1,2,5-thiadiazole show similar trends. In these molecules the H—C(3) and H—C(4) bonds are displaced several degrees from the bisectors of the corresponding ring angles, toward the adjacent nitrogen atoms.

For thiazole and pyrazole the geometries of the two sides of the rings are included in Table III. In thiazole, the geometry of the SC(2)N part of the ring resembles the corresponding part of 1,3,4-thiadiazole, while the remaining half of the ring resembles the corresponding part of thiophene.[60] Departures from this generalization are that the ring angle at the nitrogen atom is less than that at C(4), and that the H—C(4) bond is about 3° to the nitrogen side of the bisector of the ring angle. In pyrazole, the C(4)C(5)N part of the structure resembles the corresponding part of pyrrole, the bonds in pyrazole being a little shorter. The ring angle at N(2) in pyrazole is only 104.1° and the H—C(3) bond lies some 5° to the nitrogen side of the bisector of the ring angle. The N—H bond in pyrazole also does not lie on the bisector of the ring angle at N(1), as is required by symmetry in pyrrole, but is displaced from that line by some 5° toward the second nitrogen.[61,62] The tendencies for hydrogen atoms to be displaced towards hetero atoms may be connected with the presence of lone pairs of electrons on the hetero atoms, although other factors clearly also enter.[61]

The N—H substitution length in pyrrole is 0.996 Å and that in

[57] J. H. Griffiths, A. Wardley, V. E. Williams, N. L. Owen, and J. Sheridan, *Nature* (*London*) **216**, 1301 (1967).

[58] O. L. Stiefvater, H. Jones, and J. Sheridan, *Spectrochim. Acta A* **26**, 825 (1970).

[59] K. Bolton, R. D. Brown, F. R. Burden, and A. Mishra, *Chem. Commun.* 873 (1971).

[60] L. Nygaard, E. Asmussen, J. H. Høg, R. C. Maheshwari, C. H. Nielsen, I. B. Petersen, J. R. Andersen, and G. O. Sørensen, *J. Mol. Struct.* 8, 225, (1971); **9**, 220 (1971).

[61] D. Christen, L. Nygaard, and G. O. Sørensen, *Colloq. High Resolution Mol. Spectrosc.*, *Dijon* Paper F1 (1971).

[62] L. Nygaard, D. Christen, J. T. Nielsen, E. J. Pedersen, O. Snerling, E. Vestergaad, and G. O. Sørensen, *J. Mol. Struct.* (1974) (in press).

pyrazole 0.998 Å; in imidazole[63] it is also 0.998 Å. These lengths are a little shorter than the 1.02 Å found in dimethylamine, in agreement with the different bonding at nitrogen in the two types of molecule.

Microwave spectroscopy distinguishes readily between possible tautomeric forms of certain heterocyclic systems, such as the triazoles,[58,59] where the spectra are those of the asymmetric 1,2,3 and 1,2,4 forms. In tetrazole[64] two tautomers, both planar, are believed to be present.

B. Electron Distribution

The chief findings are summarized in Table IV. In those cases where the lines of action of dipole moments are not determined by symmetry, an indication of the approximate line of action is given. The ^{14}N nuclear couplings are indicated by eQq_{cc}, the C axis, perpendicular to the ring, being always a principal axis of the coupling tensor; the measured properties of the tensor in the ring plane are also indicated. Generally, for pyridine-type nitrogens, the data are consistent with a principal tensor axis, corresponding to the largest negative coupling constant, directed very roughly along the external bisector of the ring angle at the N atom. This is in accord with a similar orientation of a lone pair of electrons with some p character, but some of the data allow the small deviations of this tensor axis from the bisector to be stated. For such nitrogens, eQq_{cc} is normally positive and between 1 and 2.6 MHz, but it is less when the nitrogen is bonded to O or NH in isoxazole, 1,2,4-oxadiazole, or pyrazole. This is possibly due to ionic character in the N—O or N—NH bonds.[65] The coupling tensors for the N(2) positions in isoxazole, 1,2,4-oxadiazole, and pyrazole are very asymmetric. Pyrrole-type nitrogens have negative values of eQq_{cc}, which are smaller numerically than for a localized lone pair of electrons and reflect the π delocalization shown by the other data.

The dipole moments of some such substances substituted by methyl or CHO groups (Section IV,C) are reported and generally follow expectations. The small electronic effects of methyl groups suggest that such substitutions, which can cause large swings of principal axes, will be useful in confirming directions of dipole moments in structures where isotopic effects do not decide these directions conclusively. Such substitutions might also extend knowledge of quadrupole coupling tensors,

[63] J. H. Griffiths, Thesis, Univ. College of North Wales (1969).
[64] W. D. Krugh and L. P. Gold, *Symp. Mol. Struct. Spectrosc., Columbus, Ohio* Paper E10 (1972); *J. Mol. Spectrosc.* (1974) (in press).
[65] E. Saegebarth and A. P. Cox, *J. Chem. Phys.* **43**, 166 (1965).

TABLE IV

ELECTRON DISTRIBUTION IN AROMATIC FIVE-MEMBERED RINGS

Molecule	μ (D)	$\theta^{\circ\,a}$	$^{14}N\text{-}eQq_{cc}$ (MHz)	Other information[b]	References
 (cyclopentadiene)	0.42	0	—	$\Delta\chi = -34.2$; $\Sigma\Delta\chi_{\mathrm{gp}} = -16.7$	c–f
 (furan, O)	0.66	180	—	$\Delta\chi = -38.7$; $\Sigma\Delta\chi_{\mathrm{gp}} = -15.7$	d,f–k
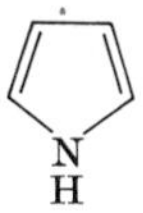 (pyrrole, N–H)	1.74	0	-2.66	$\Delta\chi = -42.4$; $\Sigma\Delta\chi_{\mathrm{gp}} = -7.9$; $eQq(\uparrow) = 1.45$; $eQq(\rightarrow) = 1.21 = eQq_\alpha$	f,j,l,m
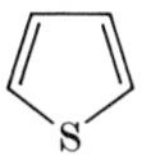 (thiophene, S)	0.55	180	—	$\Delta\chi = -50.1$; $\Sigma\Delta\chi_{\mathrm{gp}} = -16.5$	f,i,j,n
 (selenophene, Se)	0.37	180	—	$\Delta\chi = -51.0$	o–q
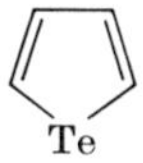 (tellurophene, Te)	0.19	180	—	—	r
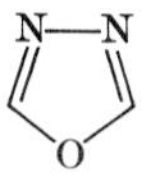 (N—N, O)	3.04	0	1.78	$eQq_\alpha = -4.83$; α axis 16° to external bisector of ring angle, toward other N	s,t
 (N—N, S)	3.28	0	2.11	$eQq(\uparrow) = -3.72$; $eQq(\rightarrow) = 1.61$	u,v
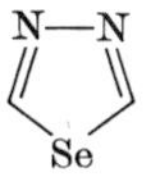 (N—N, Se)	3.40	0	—	—	w
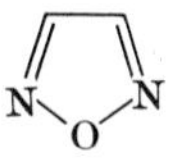 (N, O, N)	3.38	180	1.15	$eQq(\uparrow) = -0.46$; $eQq(\rightarrow) = -0.69$	x
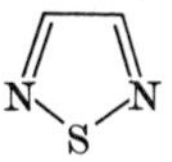 (N, S, N)	1.57	180	~ 2.6	$|eQq(\uparrow)| < 1$; $eQq(\rightarrow) = -2.6 \pm 1$	y

(continued)

Table IV (continued)

Molecule	μ (D)	 $\theta^{\circ a}$	$^{14}\text{N-}eQq_{cc}$ (MHz)	Other information[b]	References
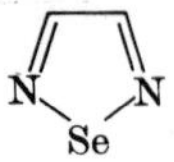 (ring: N, N, Se)	1.11	180	~ 2.6	$\lvert eQq(\uparrow)\rvert < 1$; $eQq(\rightarrow) = -2.6 \pm 1$	z
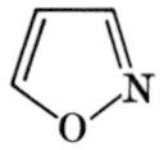 (ring: O, N)	2.90	130	~ 0.1	$eQq_\alpha \leqslant -5.3$, close to external bisector at position 2	aa–cc
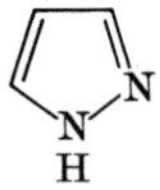 (ring: N, N, H)	2.21	69	$\begin{cases} -3.02(1) \\ 0.79(2) \end{cases}$	$eQq_\alpha(1) = 0.72$, 28° to external bisector at position 1, toward 2; $eQq_\alpha(2) = -4.48$, 18° to external bisector at position 2, toward 1	dd–ii
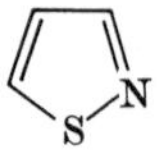 (ring: S, N)	2.4	116	1.4	$eQq_\alpha \simeq -4.7$, near external bisector at position 2	cc,jj
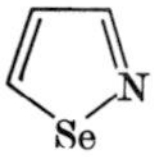 (ring: Se, N)	2.30	109	1.4	$eQq(\uparrow) = 1.1$; $eQq(\rightarrow) = -2.5$	kk
 (ring: N, O)	1.50	~ 55	2.41	$eQq_\alpha \simeq -4.2$, near external bisector at position 3	aa,cc
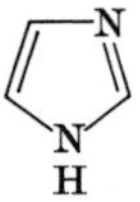 (ring: N, N, H)	3.70	17	$-2.59(1)$ \ 2.29(3)	$eQq_\alpha(1) = 1.1$, close to NN direction; $eQq_\alpha(3) = -4.08$, close to external bisector at position 3	jj,ll–nn
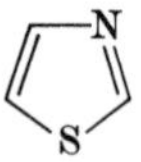 (ring: N, S)	1.61	37	2.59	$eQq_\alpha = -4.41$, close to external bisector at position 3	oo,pp
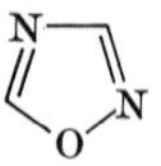 (ring: N, N, O)	1.2	~ 90	$-0.17(2)$ \ 2.1(4)	eQq_{aa} and eQq_{bb} at positions 2 and 4, consistent with tensors similar to those in isoxazole and oxazole	jj,kk,nn,qq
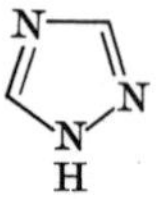 (ring: N, N, N, H)	2.72	—	—	—	rr
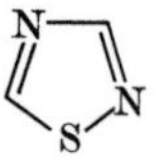 (ring: N, N, S)	1.49	62	1.3(2) \ 3.0(4)	$eQq(\uparrow)(2) = 1.05$; $eQq(\rightarrow)(2) = -2.36$; $eQq(\uparrow)(4) = -2.94$; $eQq(\rightarrow)(4) = 0.08$	nn,qq

in similar ways, in such molecules as isothiazole and isoselenazole, although it must be noted that N-methyl substitution in pyrrole[66,67] drastically changes the N coupling tensor.

In most cases, the dipole moment and nuclear coupling data can be qualitatively accounted for on the basis of bond, lone-pair, or atomic[68] moments, and valence bond theory of the type applied to quadrupole couplings by Townes and Dailey.[69] More significantly, however, SCF

[66] W. Arnold, H. Dreizler, and H. D. Rudolph, Z. Naturforsch. A 23, 301 (1968).
[67] L. Nygaard, J. T. Nielsen, J. Kirchheiner, G. Maltesen, J. R. Andersen, and G. O. Sørensen, J. Mol. Struct. 3, 491 (1969).
[68] T. D. Gierke, H. L. Tigelaar, and W. H. Flygare, J. Amer. Chem. Soc. 94, 330 (1972).
[69] C. H. Townes and B. P. Dailey, J. Chem. Phys. 17, 782 (1949).

---

[a] θ is the probable approximate angle between the direction of μ, with the sign shown, and the symmetry (or near-symmetry) axis of the ring through atom 1, in the molecular plane.

[b] $\Delta\chi$ in 10^{-6} erg G^{-2} mole^{-1}: eQq in MHz: eQq_a is most negative principal value of nuclear coupling tensor in molecular plane. $eQq(\uparrow)$ and $eQq(\rightarrow)$ indicate coupling constants referred to axes approximately parallel or perpendicular to the symmetry (or near-symmetry) axis through atoms 1, in the molecular plane. Unstated coupling constants are derivable from the data given.

[c] L. H. Scharpen and V. W. Laurie, J. Chem. Phys. 43, 2765 (1965).

[d] T. D. Gierke, H. L. Tigelaar, and W. H. Flygare, J. Amer. Chem. Soc. 94, 330 (1972).

[e] R. C. Benson and W. H. Flygare, J. Amer. Chem. Soc. 92, 7523 (1970).

[f] T. G. Schmalz, C. L. Norris, and W. H. Flygare, J. Amer. Chem. Soc. 95, 7961 (1973).

[g] M. H. Sirvetz, J. Chem. Phys. 19, 1609 (1951).

[h] B. Bak, E. Hamer, D. H. Sutter, and H. Dreizler, Z. Naturforsch. A 27, 705 (1972).

[i] D. H. Sutter and W. H. Flygare, J. Amer. Chem. Soc. 91, 4063 (1969).

[j] T. J. Barton, R. W. Roth, and J. G. Verkade, J. Amer. Chem. Soc. 94, 8854 (1972).

[k] G. R. Tomasevitch, K. D. Tucker, and P. Thaddeus, J. Chem. Phys. 59, 131 (1973).

[l] L. Nygaard, J. T. Nielsen, J. Kirchheiner, G. Maltesen, J. R. Andersen, and G. O. Sørensen, J. Mol. Struct. 3, 491 (1969).

[m] D. H. Sutter and W. H. Flygare, J. Amer. Chem. Soc. 91, 6895 (1969).

[n] T. Ogata and K. Kozima, J. Mol. Spectrosc. 42, 38 (1972).

[o] N. M. Pozdeev, O. B. Akulinin, A. A. Shapkin, and N. N. Magdesieva, Zh. Strukt. Khim. 11, 869 (1970).

[p] R. D. Brown, F. R. Burden and P. D. Godfrey, J. Mol. Spectrosc. 25, 415 (1968).

[q] W. Czieslik, D. H. Sutter, H. Dreizler, C. L. Norris, S. L. Rock, and W. H. Flygare, Z. Naturforsch. A 27, 1691 (1972).

[r] R. D. Brown and J. G. Crofts, Chem. Phys. 1, 217 (1973).

[s] L. Nygaard, R. L. Hansen, J. T. Nielsen, J. R. Andersen, G. O. Sørensen, and P. A. Steiner, J. Mol. Struct. 12, 59 (1972).

[t] B. Bak, J. T. Nielsen, O. F. Nielsen, L. Nygaard, J. R. Andersen, and P. A. Steiner, J. Mol. Spectrosc. 19, 458 (1966).

[u] B. Bak, D. Christensen, L. Nygaard, L. Lipschitz, and J. R. Andersen, J. Mol. Spectrosc. 9, 225 (1962).

MO treatments of these systems[70-75] are now showing some success in accounting for the dipole properties and electric field asymmetries at nuclei. The data in Table IV, and similar data still to be obtained, will provide good tests of theoretical descriptions of the electron distributions.

[70] D. W. Davies and W. C. Mackrodt, *Chem. Commun.* 345 (1967).
[71] D. W. Davies and W. C. Mackrodt, *Chem. Commun.* 1226 (1967).
[72] R. D. Brown and B. A. W. Coller, *Theor. Chim. Acta* **7**, 259 (1967).
[73] R. D. Brown, B. A. W. Coller, and J. E. Kent, *Theor. Chim. Acta* **10**, 435 (1968).
[74] M. Roche, F. d'Amato, and M. Benard, *J. Mol. Struct.* **9**, 183 (1971).
[75] G. L. Blackman, R. D. Brown, and F. R. Burden, *J. Mol. Spectrosc.* **36**, 528 (1970).

[v] L. Nygaard, R. L. Hansen, and G. O. Sørensen, *J. Mol. Struct.* **9**, 163 (1971).
[w] D. M. Levine, W. D. Krugh, and L. P. Gold, *J. Mol. Spectrosc.* **30**, 459 (1969).
[x] E. Saegebarth and A. P. Cox, *J. Chem. Phys.* **43**, 166 (1965).
[y] V. Dobyns and L. Pierce, *J. Amer. Chem. Soc.* **85**, 3553 (1963).
[z] G. L. Blackman, R. D. Brown, F. R. Burden, and J. E. Kent, *Chem. Phys. Lett.* **1**, 379 (1967).
[aa] W. C. Mackrodt, A. Wardley, P. A. Curnuck, N. L. Owen, and J. Sheridan, *Chem. Commun.* 692 (1966).
[bb] A. Wardley, P. Nösberger, O. L. Stiefvater, and J. Sheridan (to be published).
[cc] A. Wardley and J. Sheridan, *Eur. Microwave Spectrosc. Conf., 1st, Bangor, Wales* Paper A1–5 (1970).
[dd] W. H. Kirchhoff, *J. Amer. Chem. Soc.* **89**, 1312 (1967).
[ee] G. L. Blackman, R. D. Brown, and F. R. Burden, *J. Mol. Spectrosc.* **36**, 528 (1970).
[ff] G. L. Blackman, R. D. Brown, F. R. Burden, and A. Mishra, *J. Mol. Struct.* **9**, 465 (1971).
[gg] N. M. Pozdeev, R. S. Nasibullin, R. G. Latypova, V. G. Vinokurov, and N. D. Konevskaya, *Opt. Spektrosk.* **31**, 63 (1971).
[hh] R. S. Nasibullin, R. G. Latypova, V. S. Troitskaya, V. G. Vinokurov, and N. M. Pozdeev, *Opt. Spektrosk.* **35**, 92 (1973).
[ii] L. Nygaard, D. Christen, J. T. Nielsen, E. J. Pedersen, O. Snerling, E. Vestorgaad, and G. O. Sørensen, *J. Mol. Struct.* (1974) (in press).
[jj] J. H. Griffiths, A. Wardley, V. E. Williams, N. L. Owen, and J. Sheridan, *Nature (London)* **216**, 1301 (1967).
[kk] V. E. Williams, Theses, Univ. College of North Wales (1969); V. E. Williams and J. Sheridan (to be published).
[ll] J. H. Griffiths, Theses, Univ. College of North Wales (1969).
[mm] J. H. Griffiths, D. Christen, and J. Sheridan (to be published).
[nn] J. Sheridan, J. H. Griffiths, V. E. Williams, and D. Norbury, *Eur. Microwave Spectrosc. Conf., 1st, Bangor, Wales* Paper A1–4 (1970); V. E. Williams, D. Norbury, and J. Sheridan (to be published).
[oo] L. Nygaard, E. Asmussen, J. H. Høg, R. C. Maheshwari, C. H. Nielsen, I. B. Petersen, J. R. Andersen, and G. O. Sørensen, *J. Mol. Struct.* **8**, 225 (1971); **9**, 220 (1971).
[pp] B. Bak, D. Christensen, L. Nygaard, and J. R. Andersen, *J. Mol. Spectrosc.* **9**, 222 (1962).
[qq] D. Norbury, Thesis, Univ. College of North Wales (1974).
[rr] K. Bolton, R. D. Brown, F. R. Burden, and A. Mishra, *Chem. Commun.* 873 (1971).

Zeeman effect measurements have been made for furan,[76] thiophene,[76] pyrrole,[21] and selenophene.[77] It has not yet proved possible to determine the sign of the dipole moment in any of these substances from Zeeman studies of isotopic species, although an attempt has been made in the interesting case of furan.[78] The magnetic susceptibility anisotropies, $\Delta\chi$, and the values expected on the basis of additive group anisotropies,[22] $\Sigma\Delta\chi_{gp}$, are given in Table IV. The $\Delta\chi$ values are all much more negative than predicted, showing large nonlocal diamagnetic contributions for fields impressed perpendicular to the ring planes. This effect is also present in cyclopentadiene[79] nearly as strongly as in furan. The differences between $\Delta\chi$ and the predicted values become increasingly negative in the order: cyclopentadiene, furan, thiophene, pyrrole. This sequence is thus one of increasing aromatic character as measured by the nonlocal contributions to $\Delta\chi$. Selenophene[77] is thought to be ranked about equal to thiophene from Zeeman measurements and hence the sequence agrees well with the evidence of the bond distances (Section IV,A), although these do not indicate the same degree of similarity between furan and cyclopentadiene. The nonlocal contributions to $\Delta\chi$ in pyrrole and thiophene are almost as great as that in benzene.[22] Comparative discussions of molecular quadrupole moments, second moments of the electron distributions, and other molecular properties determined from Zeeman studies will be found in the original papers. Zeeman effects for other molecules in Table IV are likely to extend these interesting comparisons.

C. Molecular Force Fields

Centrifugal distortion constants have been measured for a number of the molecules tabulated, and some spectra of molecules in excited levels of low frequency vibrations have been studied.[65] Such data have value in deciding vibrational assignments.

Restricted rotation of side groups attached to some of these rings has been studied in a few cases. The barriers to methyl torsion in 2-methylfuran[80,81] (1190 cal mole^{-1}) and 3-methylfuran[82] (1090 cal mole^{-1}) are somewhat lower than found for methyl groups attached to

[76] D. H. Sutter and W. H. Flygare, *J. Amer. Chem. Soc.* **91**, 4063 (1969).
[77] W. Czieslik, D. H. Sutter, H. Dreizler, C. L. Norris, S. L. Rock, and W. H. Flygare, *Z. Naturforsch. A* **27**, 1691 (1972).
[78] B. Bak, E. Hamer, D. H. Sutter, and H. Dreizler, *Z. Naturforsch. A* **27**, 705 (1972).
[79] R. C. Benson and W. H. Flygare, *J. Amer. Chem. Soc.* **92**, 2 (1970).
[80] W. G. Norris and L. C. Krisher, *J. Chem. Phys.* **51**, 403 (1969).
[81] U. Andresen and H. Dreizler, *Z. Naturforsch. A* **25**, 570 (1970).
[82] T. Ogata and K. Kozima, *Bull. Chem. Soc. Japan* **44**, 2344 (1971).

vinyl-type or butadiene-type structures. In methyl-1,2,5-oxadiazole[83] (722 cal mole^{-1}), 3-methylthiophene[84,180] (740 cal mole^{-1}), and notably 2-methylthiophene[85] (555 cal mole^{-1}) they are still lower. In N-methylpyrrole[66] symmetry requires a zero threefold barrier and the sixfold barrier is only about 130 cal mole^{-1}, although this is larger than other sixfold barriers. In N-methylimidazole[86] it is not surprising to find that the geometric changes associated with introduction of a new hetero atom, even in another part of the molecule, have destroyed the symmetry sufficiently to give rise to a small threefold barrier. Several other methyl derivatives of aromatic five-membered heterocycles are under investigation and may lead to more systematic understanding of the factors affecting torsion barriers, and also to further knowledge of electron distributions (Section IV,B).

In furan-2-aldehyde (furfural)[87,88] the CHO group lies, as expected, in the plane of the ring, and the two isomers in which the carbonyl group is *cis* and *trans*, respectively, to the ring oxygen have been studied. From spectra of torsionally excited molecules, an approximate potential function for *cis–trans* conversion is derived. The barrier to conversion is about 9 kcal mole^{-1}. Intensities of spectra show the *trans* isomer to be the more stable by about 720 cal mole^{-1}. In thiophene-2-aldehyde,[87] measurements have been made on what is thought to be the isomer with the carbonyl *trans* to the sulfur.

A study has been made of 3,4-*benz* derivatives of 1,2,5-oxadiazole and 1,2,5-thiadiazole.[89] Although these structures are too large for complete geometrical determination, it is interesting that the data are best interpreted in terms of a certain amount of bond fixation in favor of the *o*-quinonoid forms of the benzene rings which place double bonds in the positions of higher bond order in the five-membered rings.

V. PARTIALLY AND FULLY SATURATED FIVE-MEMBERED RINGS

A. Molecular Geometry

As with four-membered rings, interest has centred chiefly on the planarity or otherwise of the rings and ring-puckering vibrations (Section V,C). Detailed molecular geometries have rarely been determined

[83] E. Saegebarth, *J. Chem. Phys.* **52**, 1476 (1970).

[84] T. Ogata and K. Kozima, *J. Mol. Spectrosc.* **42**, 38 (1972).

[85] N. M. Pozdeev, L. N. Gunderova, and A. A. Shapkin, *Opt. Spektrosk.* **28**, 254 (1970).

[86] H.-U. Wenger and J. Sheridan, *Eur. Microwave Spectrosc. Conf., 2nd, Bangor, Wales* Paper 1A5 (1972) (to be published).

[87] F. Mönnig, H. Dreizler and H. D. Rudolph, *Z. Naturforsch. A* **20**, 1323 (1965).

[88] F. Mönnig, H. Dreizler and H. D. Rudolph, *Z. Naturforsch. A* **21**, 1633 (1966).

[89] N. M. D. Brown, D. G. Lister and J. K. Tyler, *Spectrochim. Acta A* **26**, 2133 (1970).

by the microwave method in these cases. Accurate bond lengths and angles have been found for the planar ring in vinylene carbonate, O—CH=CH—O—C=O.[90] The C=C bond (1.331 Å) has its normal length and the ring C—O bonds also follow expectations. The carbonyl C=O length (1.191 Å) is a little shorter than in ketones and esters. Ring bond distances and angles, conforming with expectations, have been found for 1,2,4-trioxacyclopentane (ethylene ozonide).[91]

B. Electron Distribution

Dipole moments have been determined for a few such substances. The values for 2,3-dihydrofuran[92] [1.32 D probably directed roughly parallel to the C(4)–0 line], 2,5-dihydrofuran[93] (1.59 D), and tetrahydrofuran[94] (1.75 D) may be compared with the moment of furan (0.66 D) and that of cyclopentene[95] (0.22 D). Other moments include tetrahydrothiophene[96] (1.89 D), tetrahydroselenophene[97] (1.93D), germacyclopentane[98] (0.67 D), 1,2,4-trioxacyclopentane (ethylene ozonide)[91] (1.09 D), 1,2,4-trioxa-3,5-diborolane[99] (O—BH—O—BH—O) (0.96 D), 1,3,2-dioxaborolane[100] (O—CH$_2$—CH$_2$—O—BH) (2.28 D), vinylene carbonate[101] (4.57 D), maleic anhydride[102] (3.95 D), and γ-butyrolactone[103] (4.27 D).

Few nuclear quadrupole coupling cases are met, but the coupling constants for ^{73}Ge in germacyclopentane[98] and for ^{11}B in 1,3,2-dioxaborolane[100] have been measured and discussed.

Zeeman effect measurements have been made on vinylene carbonate

[90] W. F. White and J. E. Boggs, *J. Chem. Phys.* **54**, 4714 (1971).

[91] C. W. Gillies and R. L. Kuczkowski, *J. Amer. Chem. Soc.* **94**, 6337, 7609 (1972).

[92] J. R. Durig, Y. S. Li and C. K. Tong, *J. Chem. Phys.* **56**, 5692 (1972).

[93] G. G. Engerholm, Dissertation, Univ. California, 1965; *Diss. Abstr.* **26**, 7060 (1966).

[94] G. G. Engerholm, A. C. Luntz, W. D. Gwinn, and D. O. Harris, *J. Chem. Phys.* **50**, 2446 (1969).

[95] G. W. Rathjens, *J. Chem. Phys.* **36**, 2401 (1962).

[96] N. M. Pozdeev and C. C. Costain, *Trans. Comm. Spectrosc. Acad. Sci. USSR* **3**, 231 (1965); A. Kh. Mamleev and N. M. Pozdeev, *Zh. Strukt. Khim.* **10**, 747 (1969).

[97] A. Kh. Mamleev, N. N. Magesieva, and N. M. Pozdeev, *Zh. Strukt. Khim.* **11**, 1124 (1970).

[98] E. C. Thomas and V. W. Laurie, *J. Chem. Phys.* **51**, 4327 (1969).

[99] W. V. F. Brooks, C. C. Costain, and R. F. Porter, *J. Chem. Phys.* **47**, 4186 (1967).

[100] J. H. Hand and R. H. Schwendeman, *J. Chem. Phys.* **45**, 3349 (1966).

[101] K. L. Dorris, C. O. Britt, and J. E. Boggs, *J. Chem. Phys.* **44**, 1352 (1966).

[102] V. Williams, J. Sheridan, I. A. Ramsay, and L. C. Krisher, *Symp. Mol. Struct. Spectrosc., Columbus, Ohio* Paper G11 (1967) (to be published).

[103] J. R. Durig, Y. S. Li, and C. K. Tong, *J. Mol. Struct.* **18**, 269 (1973).

and maleic anhydride.[104] The magnetic susceptibility anisotropies, like those of both cyclopentenones,[105] are predictable from local group anisotropies. Nonlocal effects, such as aromaticity, are thus not indicated.

C. Molecular Force Fields

When the number of ring atoms exceeds four, there is more than one mode of nonplanar ring vibration, and in the case of five-membered rings two such vibrations occur, in which the ring atoms undergo displacement motions from a totally planar ring conformation in the sense indicated in Fig. 1. The usual + and − sign convention indicates

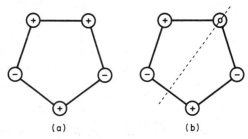

(a)   (b)

FIG. 1. Out-of-plane atomic displacements in five-membered rings; (a) is the bending vibration (species B) and (b) is the twisting vibration (species A). When equilibrium atomic positions have finite out-of-plane displacements represented by the same convention, (a) is the bent, or envelope, form (symmetry C_S), and (b) is the twisted, or half-chair, form (symmetry C_2, axis of symmetry shown dotted). Passage from form (a) to form (b) occurs by returning the upper right atom in (a) to zero displacement, and continuation of this movement produces another bent form; cyclic reversal of signs of displacement in (a) leads, in five steps, to the inverted form of (a) and in five more steps to the original form (a).

displacements to one or other side of the reference plane, while the atom with the zero sign remains in that plane. These vibrations are known as the bending and twisting vibrations, respectively. Each may have a double-minimum energy-displacement function, and, if so, we have the corresponding "bent" or "twisted" equilibrium conformations. Since both bending and twisting may reduce repulsion of any methylene hydrogens at the expense of only small increases in the ring strain, either or both vibrations are low frequency motions of large amplitude. If, as occurs in certain fully saturated five-membered rings, the two vibrations are of similar frequency, they can interact to produce the

[104] R. C. Benson and W. H. Flygare, *J. Chem. Phys.* **58**, 2366 (1973).
[105] C. L. Norris, R. C. Benson, and W. H. Flygare, *Chem. Phys. Lett.* **10**, 75 (1971).

phenomenon of pseudorotation[51,52,106] in which an out-of-plane atomic displacement may be considered to move round the ring, the conformation passing successively through ten bent and ten twisted forms in returning to its initial condition. The totally planar state may have much higher energy than any conformation adopted during the above process, and therefore different conformations can be interconverted without the requirement of the energy which may be necessary to make the ring planar.

Pseudorotation is an important property of cyclopentane, in which all ten bent conformations traversed in each pseudorotational cycle are equivalent and differ little in energy from the ten equivalent twisted conformations which are also traversed. Cyclopentane is known from vibrational spectra,[107] and other evidence, to undergo virtually free pseudorotation, the energy of the ring remaining essentially constant during the pseudorotational cycle.

When, as in all molecules suitable for microwave study, the five ring groups are not all equivalent, the pseudorotation will involve passage through conformations of different energy, and the most energetically favorable ones will be separated by barriers to restricted pseudorotation, which vary greatly in magnitude. A particularly valuable discussion of these properties of five-membered rings in relation to microwave spectroscopy is given by Harris et al.[108]

When one ring bond is a double bond, torsion at that bond during pseudorotation requires much energy, in contrast with the relief of H⋯H or similar repulsions when such torsion occurs at a single ring-bond. Such substances accordingly favor the bent conformation, in which the adjacent ring atoms which have the same sign of displacement are those doubly bonded; pseudorotation is heavily restricted. In simpler terms, the presence of the double bond makes the twisting frequency much higher than the bending frequency, and these two modes can accordingly be treated independently. The preferred bent conformation undergoes ring flexing in a way similar to that of a four-ring compound, the doubly bonded atoms acting as one in this vibration. Such cases are conveniently considered first. These and several similar substances suitable for microwave study have been extensively investigated in complementary ways by vibrational spectroscopy, and microwave studies of further members of this class are likely.

[106] J. E. Kilpatrick, K. S. Pitzer, and R. Spitzer, J. Amer. Chem. Soc. 69, 2483 (1947).
[107] J. R. Durig and D. W. Wertz, J. Chem. Phys. 49, 2118 (1968).
[108] D. O. Harris, G. G. Engerholm, C. A. Tolman, A. C. Luntz, R. A. Keller, H. Kim, and W. D. Gwinn, J. Chem. Phys. 50, 2348 (1969).

1. *Five-Membered Rings Containing One Endocyclic Double Bond*

The homocyclic reference substance, cyclopentene,[109,110] has the bent geometry discussed above, with the central methylene carbon out of the plane of the other four carbons. Inversion of the ring is opposed by a barrier of 663 cal mole^{-1}. The two lowest levels of the ring bending mode are separated by only 0.91 cm^{-1}, in contrast with the twisting frequency which occurs at 386 cm^{-1}. As with cyclobutane, replacement of a CH_2 by carbonyl greatly reduces the barrier to ring inversion, and both the cyclopentenones[111,112] have planar equilibrium geometry.

In 2,3-dihydrofuran,[92] the replacement of one methylene group in cyclopentene by oxygen leads to a lowering of the barrier to ring inversion to about 237 cal mole^{-1}. In 2,5-dihydrofuran, where the remaining methylene groups are now separated by the oxygen atom, it is not surprising to find that such a barrier is no longer present, the molecule showing a planar ring conformation of minimum energy.[93] When one or both of the methylene groups in 2,5-dihydrofuran is replaced by carbonyl groups to form γ-crotonolactone[113] and maleic anhydride,[102] respectively, the rings are planar, as is that in vinylene carbonate.[101]

The results for dihydrothiophenes are similar. The 2,3 isomer shows nonplanarity with a moderate barrier to ring inversion, while the 2,5 isomer has a planar conformation.[114]

As in the case of four-membered rings, the potential energy for the ring-bending vibrations can be expressed as a sum of terms which are quartic and quadratic in the displacement coordinate. Negative quadratic terms are again associated with double minimum potentials and nonplanar equilibrium configurations, while positive quadratic terms occur when the rings are planar. Most details of this type for this group of substances are found in the vibrational studies.[115–117]

A microwave study of 2,5-dihydropyrrole[118] accords with vibra-

[109] S. S. Butcher and C. C. Costain, *J. Mol. Spectrosc.* **15**, 40 (1965).
[110] L. H. Scharpen, *J. Chem. Phys.* **48**, 3552 (1968).
[111] D. Chadwick, A. C. Legon, and D. J. Millen, *Chem. Commun.* 1130 (1969).
[112] J. W. Bevan and A. C. Legon, *J. Chem. Soc. Faraday Trans. II* **69**, 902 (1973).
[113] A. C. Legon, *Chem. Commun.* 838 (1970).
[114] J. A. Greenhouse, *Symp. Mol. Struct. Spectrosc., Columbus, Ohio* Paper X4 (1968); and private communication.
[115] T. Ueda and T. Shimanouchi, *J. Chem. Phys.* **47**, 4042, 5018 (1967).
[116] W. H. Green and A. B. Harvey, *J. Chem. Phys.* **49**, 177 (1968).
[117] L. A. Carreira and R. C. Lord, *J. Chem. Phys.* **51**, 3225 (1969).
[118] C. R. Nave and K. P. Pullen, *Chem. Phys. Lett.* **12**, 499 (1972).

tional data[119,120] which indicate nonplanarity and an asymmetric double minimum potential on account of the axial–equatorial conversion of the N—H bond by ring inversion. Probably the equatorial form is favored energetically.

We may also include in this section studies of 2-epoxy derivatives of singly unsaturated five-membered rings. When an endocyclic double bond is replaced by a three-membered ring, since the two rings are *cis* to each other, the bent form of the five-membered ring will clearly remain preferred, but two arrangements will be possible depending on whether the bending occurs toward or away from the three-membered ring. In cyclopentene oxide,[119,121] the moments of inertia and dipole direction show that the central methylene of the five-membered ring is *cis* to the oxygen, in resemblance to the boat configuration of cyclohexane. The inertial properties of 3,6-dioxabicyclo(3.1.0)hexane,[122] the epoxy derivative of 2,5-dihydrofuran, indicate a similar conformation, as does the fact that the dipole moment of 2.49 D is directed along a line almost perpendicular to the plane of the four carbon atoms. The difference between this case and that of 2,5-dihydrofuran follows expectations based on the different repulsive forces on the methylene groups in the epoxy derivative.

2. Saturated Five-Membered Rings

When the ring groups are not all alike in such substances, the degree to which a particular ring conformation is preferred, and the heights of barriers to pseudorotation, appear sensitive to a number of factors which vary considerably from molecule to molecule. It must again be noted that much of our knowledge of such systems has come from complementary studies by both microwave and vibrational spectroscopy, and only an occasional reference to vibrational spectra can be made here.

A particularly detailed study of tetrahydrofuran has been made by the microwave method.[94] Spectra were assigned for the ground state and no fewer than eight excited states with levels below 200 cm^{-1}. Analysis leads to a detailed potential function for pseudorotation[108,123] which is opposed by a barrier of only some 160 cal mole^{-1}. The variation of dipole properties with the pseudorotational energy level indicates a twisted configuration as the more stable form. Low frequency

[119] L. A. Carreira and R. C. Lord, *J. Chem. Phys.* **51**, 2735 (1969).

[120] L. A. Carreira, R. O. Carter, and J. R. Durig, *J. Chem. Phys.* **57**, 3384 (1972).

[121] W. J. Lafferty, *J. Mol. Spectrosc.* **36**, 84 (1970).

[122] R. A. Creswell and W. J. Lafferty, *J. Mol. Spectrosc.* **46**, 371 (1973).

[123] R. Davidson and P. A. Warsop, *J. Chem. Soc. Faraday Trans. II* **68**, 1875 (1972).

infrared spectra confirm the findings.[124] The low energy requirement for interconversion of the two forms through pseudorotation may be compared with the energy necessary for interconversion via the planar configuration, which is 3500 cal mole^{-1}. Detailed knowledge of such low-energy paths for changes in ring conformation are obviously important in many chemical situations.

In 1,3-dioxolane, in which a further methylene group is replaced by oxygen, the situation is similar to that in tetrahydrofuran,[123–125] with a low barrier to pseudorotation; in this case the more stable form appears to be a bent one.

For 1,2,4-trioxacyclopentane (ethylene ozonide) the equilibrium conformation is twisted,[91] and there is a barrier of at least some 1500 cal mole^{-1} opposing pseudorotation. This configuration and large barrier may well result from opposition to torsion about the O—O bond, e.g., by lone pair repulsions, in a way not met in the previously mentioned examples. Vibration satellite spectra are shown conclusively, by nuclear statistical effects, to be due to molecules excited in the ring bending mode, which has a frequency of some 200 cm^{-1}.

In γ-butyrolactone,[103,113] the ring is nonplanar and there is a dipole component of 0.33 D roughly perpendicular to the ring. This introduction of a carbonyl group into tetrahydrofuran leads to independent ring bending and twisting modes at 148 and 225 cm^{-1}, respectively. The same effect is observed when a carbonyl group is substituted into cyclopentane, cyclopentanone[126] being permanently twisted with independent bending and twisting modes.

Ethylene carbonate[126a] has inertial properties and vibrational satellite spectra showing a nonplanar equilibrium conformation of the ring.

Microwave studies of tetrahydrothiophene[96] and tetrahydroselenophene[97] are interpreted in terms of a twisted equilibrium conformation. Vibrational studies[127,127a] indicate barriers to pseudorotation which are much higher than in tetrahydrofuran. Electron diffraction results[128,129] for the sulfur and selenium compounds, as with some other examples, support the findings.

[124] J. A. Greenhouse and H. L. Strauss, *J. Chem. Phys.* **50**, 124 (1969).

[125] D. O. Harris and P. A. Baron, private communication.

[126] H. Kim and W. D. Gwinn, *J. Chem. Phys.* **51**, 1815 (1969).

[126a] I. Wang, C. O. Britt and J. E. Boggs, *J. Amer. Chem. Soc.* **87**, 4950 (1965).

[127] D. W. Wertz, *J. Chem. Phys.* **51**, 2133 (1969).

[127a] W. H. Green, A. B. Harvey, and J. A. Greenhouse, *J. Chem. Phys.* **54**, 850 (1971).

[128] Z. Nahlovska, B. Nahlovsky, and H. M. Seip, *Acta Chem. Scand.* **23**, 3534 (1969).

[129] Z. Nahlovska, B. Nahlovsky, and H. M. Seip, *Acta Chem. Scand.* **24**, 1903 (1970).

The twisted conformation is again found to be preferred in germa-cyclopentane,[98] where microwave data indicate a barrier of at least 1400 cal mole^{-1} opposing pseudorotation; vibrational spectra[130] indicate a barrier of nearly 6 kcal mole^{-1}. A similar situation occurs in silacyclopentane.[131,132]

Two further rings, containing boron, have been similarly studied. Data for 1,3,2-dioxaborolane[100] O—CH$_2$—CH$_2$—O—BH are interpreted in terms of a slightly twisted equilibrium ring geometry, with a small barrier to inversion. The ring is planar in 1,2,4-trioxa-3,5-diborolane[99] O—BH—O—BH—O, although ring deformations appear considerably anharmonic.

Two methyl derivatives of 1,2,4-trioxacyclopentane, propylene ozonide, and trans-2-butylene ozonide,[133] have been investigated. Rotational constants, including those for some isotopically enriched species, and dipole properties, show the spectra of these substances to be caused by twisted conformations, as ethylene ozonide, with methyl groups at equatorial positions. Microwave findings for ozonides[91,133] have important bearings on the mechanism of ozonolysis.

VI. AROMATIC SIX-MEMBERED RINGS

A. Molecular Geometry

A particularly detailed study has been made of pyridine[134,135,139] leading to a complete substitution geometry. All the C—C bonds are close to 1.393 Å in length, only some 0.004 Å shorter than those in benzene. The shorter C—N bonds (1.338 Å) lead to a shorter distance across the ring between the 2- and 6-positions than between the 3- and 5-positions; the ring angle at the nitrogen (116.9°) and that at

[130] J. R. Durig and J. N. Willis, J. Chem. Phys. **52**, 6108 (1970); see also J. R. Durig, Y. S. Li, and L. A. Carreira, J. Chem. Phys. **58**, 2393 (1973).

[131] J. R. Durig and J. N. Willis, J. Mol. Spectrosc. **32**, 320 (1969).

[132] W. J. Lafferty, private communication.

[133] R. P. Lattimer, C. W. Gillies, and R. L. Kuczkowski, J. Amer. Chem. Soc. **95**, 1348 (1973).

[134] G. O. Sørensen, J. Mol. Spectrosc. **22**, 325 (1967).

[135] B. Bak, L. Nygaard, and J. R. Andersen, J. Mol. Spectrosc. **2**, 361 (1958); see also G. O. Sørensen, L. Nygaard, and L. Mahler, Eur. Mol. Spectrosc. Congr., 8th, Copenhagen Paper 362 (1965).

[136] W. Werner, H. Dreizler, and H. D. Rudolph, Z. Naturforsch. A **22**, 531 (1967).

[137] G. L. Blackman, R. D. Brown, and F. R. Burden, J. Mol. Spectrosc. **35**, 444 (1970).

[138] R. L. Kuczkowski and A. J. Ashe III, J. Mol. Spectrosc. **42**, 457 (1972).

[139] G. O. Sørensen, L. Mahler, and N. Rastrup-Andersen, J. Mol. Struct. (1974) (in press).

C(3) (118.5°) and C(4) (118.4°) are all slightly less than the 120° value in benzene, the ring angle at C(2) being correspondingly greater at 123.8°. The C—H lengths are all near 1.083 Å. The C—H bond at the 4-position naturally lies on the symmetry axis and bisects the ring angle exactly, while that at the 3-position also lies within 1° of the bisector of the ring angle; the C—H bond at the 2-position lies some 2° to the nitrogen side of the ring angle bisector, a finding analogous to that for similar C—H bonds in aromatic five-membered rings.

Similar but less complete studies have been made for pyridazine[136] and pyrimidine.[137] In the first of these, the B axis passes close to the C and H atoms at the 3- and 6-positions, and these atoms were not accurately located. The C—C bond joining the 4- and 5-positions is found to be 1.375 Å, somewhat shorter than such bonds in pyridine, while the N—N bond is found to be 1.330 Å. In pyrimidine, the inertial axes should allow a complete structure determination, and we may expect accurate findings for this and related structures, as well as of their simple substitution products.

Phosphabenzene, the phosphorus analogue of pyridine, has recently been studied.[138] It is a planar molecule, and the data agree with a P—C distance of about 1.72 Å and a phosphorus ring angle of about 103°.

B. Electron Distribution

Dipole moments have been determined for pyridine[139] (2.22 D), pyridazine[136] (4.22 D), pyrimidine[137] (2.33 D), and phosphabenzene[138] (1.54 D). In pyridine-N-oxide[140] the dipole moment (4.13 D) has been compared with predictions from molecular orbital theory (3.8–6.9 D). A similar comparison is made for the magnitude and direction of the moment in 2-fluoropyridine[141] (3.40 D) and an INDO calculation, which gave good agreement. The dipole moments in 2-methyl-[142] and 4-methylpyridine[143] have about the expected magnitudes and orientations.

Nitrogen nuclear quadrupole couplings have been measured for a number of these substances. In pyridine and pyridazine, the principal values and axes of the coupling tensors have been determined. The coupling constants with respect to axes perpendicular to the rings are positive (~ 3 MHz), and the largest negative value of the coupling

[140] R. D. Brown, F. R. Burden, and W. Garland, *Chem. Phys. Lett.* **7**, 461 (1970).

[141] S. D. Sharma, S. Doraiswamy, H. Legell, H. Mäder, and D. H. Sutter, *Z. Naturforsch.* *A* **26**, 1342 (1971).

[142] H. Dreizler, H. D. Rudolph, and H. Mäder, *Z. Naturforsch.* *A* **25**, 25 (1970).

[143] H. D. Rudolph, H. Dreizler, and H. Seiler, *Z. Naturforsch.* *A* **22**, 1738 (1967).

constant lies near the external bisector of the ring angle at N, as expected from the bonding. In less well-determined cases, the coupling constants accord with similar properties of the coupling tensors, which resemble those for pyridine-like nitrogen atoms in aromatic five-membered rings. The ^{14}N couplings in 2-fluoropyridine[141] have been considered in detail. Pentafluoropyridine[181] has been studied.

Nuclear coupling constants have been measured for bromine in 2- and 4-bromopyridines[144] and for chlorine in 2-chloropyridine,[145] and the findings analyzed in terms of a few percent double bond character in the halogen–carbon bonds (cf. new data for 2-chlorothiophene[182]).

Zeeman effect studies have been made for pyridine[146] and 2-fluoropyridine.[147] The magnetic susceptibility anisotropy, $\Delta\chi$, for pyridine (-57.4×10^{-6} erg G^{-2} mole^{-1}) is nearly as great as that in benzene and corresponds to a large nonlocal diamagnetism to fields impressed perpendicular to the ring plane and to high aromatic character. In 2-fluoropyridine, $\Delta\chi$ is similar (-52.1×10^{-6} erg G^{-2} mole^{-1}).

In a recent study,[148] the Zeeman effect for different isotopic forms of pyridine indicates the sign of the dipole moment with the negative end of the dipole toward the nitrogen atom.

C. Molecular Force Fields

Barriers to internal rotation of methyl groups have been studied for 2-[142] and 4-methylpyridines.[143] In the 4-methyl compound, symmetry requires a zero threefold barrier, and the sixfold barrier is only 13.51 cal mole^{-1}. In 2-methylpyridine the loss of symmetry allows a threefold barrier, the most probable value of which is 258.4 cal mole^{-1}, with an additional sixfold barrier of opposite sign, -11.8 cal mole^{-1}.

The spectrum of 2-aminopyridine[149] shows the effects of the inversion of the amino group which are found in the microwave spectrum of aniline[150] and other amines and amides. The NH_2 plane is some 32° out of the plane of the remainder of the molecule, and the two lowest energy levels in the double minimum inversion mode are separated by

[144] W. Caminati and P. Forti, *Chem. Phys. Lett.* **15**, 343 (1972).

[145] F. Scappini and A. Guarnieri, *Z. Naturforsch. A* **27**, 1011 (1972).

[146] J. H. S. Wang and W. H. Flygare, *J. Chem. Phys.* **52**, 5636 (1970).

[147] D. H. Sutter, *Z. Naturforsch. A* **26**, 1644 (1971).

[148] E. Hamer and D. H. Sutter, *Colloq. High Resolution Mol. Spectrosc., Dijon* Paper D6 (1971); D. H. Sutter, E. Hamer, W. Czieslik, and L. Engelbrecht, *Eur. Microwave Spectrosc. Conf., 2nd, Bangor, Wales* Paper 3A1 (1972).

[149] R. A. Kydd and I. M. Mills, *J. Mol. Spectrosc.* **42**, 320 (1972).

[150] D. G. Lister and J. K. Tyler, *Chem. Commun.* 152 (1966).

about 135 cm^{-1}. Similar studies on 3- and 4-aminopyridine are being made, with comparable findings.[151]

Studies of other substituted pyridines with different possible conformations of substituents are likely.

VII. PARTIALLY AND FULLY SATURATED SIX-MEMBERED RINGS

These substances in general show definite preference for certain ring geometries, and pseudorotation is not normally encountered. As with the five-membered rings, it is convenient to classify molecules into the partially and fully saturated categories.

A. Six-Membered Rings Containing Endocyclic Double Bonds

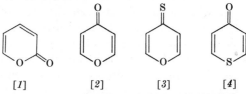

[1]   [2]   [3]   [4]

Two endocyclic double bonds occur in 2- [1] and 4-pyrone [2], and they are in conjugation with a carbonyl group in each case. Tyler and co-workers[152,153] have shown that 4-pyrone is a planar molecule and have obtained a complete structure by extensive isotopic substitution. The distances suggest a good deal of electron delocalization in the ring. The ring C—O bond (1.355 Å) is shorter than for a single bond, being even a little shorter than in furan, while the formal C=C bonds (1.356 Å) are almost as long as those in furan. The formal C—C bonds (1.440 Å) are shorter than the central bond in butadiene, although in pyrone they will be influenced by the carbonyl group. The carbonyl bond itself (1.248 Å) is noticeably long. The C—H bonds have their expected lengths; the C(3)—H(3) bond is within 1° of the ring angle bisector at C(3), but once again the C(2)—H(2) bond is tilted toward the hetero atom, here by some 7° from the ring angle bisector.

The 2-pyrone molecule is also planar,[154] as are pyran-4-thione [3][152,153] and thiapyran-4-one [4].[152,155] In pyran-4-thione the moments of

[151] D. G. Lister, private communication.

[152] J. K. Tyler, private communication (to be published); Eur. Microwave Spectrosc. Conf., 1st Bangor, Wales Paper B2–4 (1970).

[153] J. N. Macdonald, Thesis, Univ. Glasgow (1969).

[154] C. L. Norris, R. C. Benson, P. Beak, and W. H. Flygare, J. Amer. Chem. Soc. 95, 2766 (1973).

[155] S. Mackay, Thesis, Univ. Glasgow (1971).

inertia of several isotopic forms suggest a model with, not unexpectedly, less delocalization in the conjugated part of the molecule than in 4-pyrzone. The related 1,3-cyclohexadiene[156,157] has a nonplanar ring, although the evidence is less conclusive for 1,4-cyclohexadiene.[158] Overall, the planarity of [1]–[4] is further evidence of delocalization.

Six-membered rings with one endocyclic double bond, or cis-fused to a three-membered ring, have chiefly been studied for ring conformations. Microwave spectra of 2,3-dihydropyran[159] [5] accord with a twisted conformation in which two of the remaining methylene groups are twisted with respect to the approximately planar CH_2—CH=CH—O structure, a geometry implied by infrared studies[160] and similar to that

[5]

found for cyclohexene.[161,162] The moments of inertia of three exacted levels of a ring bending mode of 2,3-dihydropyran show little evidence of anharmonicity.[159]

Cyclohexene oxide[163] and cyclohexene sulfide[164] show microwave spectra consistent with twisted structures analogous to that for cyclohexene, the three-membered fused rings replacing the C=C double bond.

The dipole moment of 4-pyrone[152,153] has been measured as 3.70 D from Stark shifts of frequency which depart from the usual proportionality to the square of the applied field. The moment of pyran-4-thione[152,153] is 3.85 D. These moments and those measured in other studies mentioned are in general accord with the geometries and structures. The moment of 2,3-dihydropyran[159] (1.37 D) probably has a small component in a direction roughly perpendicular to the ring, which supports the nonplanar conformation.

[156] S. S. Butcher, J. Chem. Phys. 42, 1830 (1966).
[157] G. Dallinga and L. H. Toneman, J. Mol. Struct. 1, 11 (1967).
[158] G. Dallinga and L. H. Toneman, J. Mol. Struct. 1, 117 (1967).
[159] V. E. Williams, Thesis, Univ. College of North Wales (1969); V. E. Williams and J. Sheridan (to be published).
[160] R. C. Lord, T. C. Rounds, and T. Ueda, J. Chem. Phys. 57, 2572 (1972).
[161] L. H. Scharpen, J. E. Wollrab, and D. P. Ames, J. Chem. Phys. 49, 2368 (1968).
[162] J. F. Chiang and S. H. Bauer, J. Amer. Chem. Soc. 91, 1898 (1969).
[163] T. Ikeda, R. Kewley, and R. F. Curl, J. Mol. Spectrosc. 44, 459 (1972).
[164] R. Kewley, Can. J. Chem. 51, 529 (1973).

Zeeman effects in pyrones are important in relation to electron delocalization and possible aromatic character. The magnetic susceptibility anisotropies, $\Delta\chi$ (in 10^{-6} erg G^{-2} mole^{-1}), are -24.8 and -22.9 for 2- and 4-pyrone, respectively[154,164a] and are slightly less negative than values predicted from local group anisotropies. The absence of nonlocal contributions indicates that, by this criterion, the pyrones are nonaromatic. Although bond lengths in 4-pyrone suggest some delocalization of electrons, this is apparently not "cyclic" in the sense of allowing ring currents.

B. Saturated Six-Membered Rings

In the few examples studied, the species detected have chair-type configurations, except in a few cases where ring fusion dictates otherwise. In tetrahydropyran (oxacyclohexane)[165] the moments of inertia and orientation of the dipole moment of 1.58 D definitely indicate the chair geometry, which is also consistent with similar data for 1,3-dioxane.[166] Similar data show the same conformation in thioxane, $S(CH_2CH_2)_2O$, and pentamethylene sulfide.[132]

1,3,5-Trioxane also has the chair configuration and is one of the few heterocyclic molecules in the class of symmetric rotors. The threefold symmetry axis, in which lies the dipole moment of 2.07 D,[167] is consistent only with the chair form. Ring distances and angles have been estimated from isotopic substitution effects at C and O[167,168] and are normal. Excited vibrational states of trioxane have been studied in great detail through their microwave transitions, and rotational quantum numbers above 20 have been observed at very high microwave frequencies.[169,170] The assignment of these data in terms of rotation–vibration theory has been considered at length in these and other publications by the Lille group.

Paraldehyde, 2,4,6-trimethyltrioxane, also contains the symmetric rotor in which the methyl groups all occupy equatorial positions in the chair conformation.[171]

164a R. C. Benson, C. L. Norris, W. H. Flygare, and P. Beak, J. Amer. Chem. Soc. **93**, 5591 (1971).

165 V. M. Rao and R. Kewley, Can. J. Chem. **47**, 1289 (1969); **50**, 955 (1972).

166 R. Kewley, Can. J. Chem. **50**, 1690 (1972).

167 T. Oka, K. Tsuchiya, S. Iwata, and Y. Morino, Bull. Chem. Soc. Japan **37**, 4 (1964).

168 J. M. Colmont and J. C. Depannemaecker, Eur. Microwave Spectrosc. Conf., 2nd, Bangor, Wales Paper 2B-1 (1972).

169 J. Bellet, J. M. Colmont, and J. Lemaire, J. Mol. Spectrosc. **34**, 190 (1970).

170 J. M. Colmont, J. C. Depannemaecker, and J. Bellet, Eur. Microwave Specrosc. Conf., 1st, Bangor, Wales Paper E2–6 (1970).

171 R. Kewley, Can. J. Chem. **48**, 852 (1970).

When NH groups are present in similar rings, isomerism can arise depending on whether the NH bond is axial or equatorial. In piperidine,[172] both equatorial and axial forms of the chair conformation are present, clearly indicated by the isotopic shifts when NH is converted to ND. The NH-equatorial conformer is stabler by very roughly 240 cal mole^{-1}. The axial conformer has two vibrations at 240 cm^{-1} which differ in frequency by only 0.15 cm^{-1} and show strong Coriolis interaction.

The spectra observed for morpholine,[173]

$$\underbrace{NH—CH_2—CH_2—O—CH_2—CH_2}$$

are those of the chair form with equatorial NH, as shown clearly by the isotopic shift in the ND molecule and confirmed by the ^{14}N nuclear couplings and the dipole moment of 1.55 D, of which only a small component is perpendicular to the plane about which the ring atoms are distributed. A similar geometry is found for N-methyl morpholine[174]; the barrier to methyl rotation is some 1650 cal mole^{-1}.

Several "cage" molecules containing fused rings have been studied by microwave spectroscopy. Quinuclidine (1-azabicyclo[2.2.2]octane)[175] $N(CH_2CH_2)_3CH$ is constrained by its ring fusion to contain the boat conformation of the piperidine rings, and the twisting about the $CH_2—CH_2$ bonds which would relieve H$\cdots$H repulsions is opposed by increasing ring strain in a way similar to that met in saturated four- and five-membered rings. Spectra of the eight lowest states of the skeleton twisting mode show that its potential function is almost independent of the angle of twist, ϕ, between $\phi \simeq +12°$ and $\phi \simeq -12°$; there is a very small barrier of about 40 cal mole^{-1} at $\phi = 0$, the idealized boat conformation, although this is the average geometry in all states, over the torsional vibration. The force field in 1-halogen derivatives of bicyclo[2.2.2]octane[176] is similar and study of related cage systems may illustrate the balance between repulsive forces and ring strain in ways analogous to those for simple rings.

Cages of the adamantane type contain the chair conformation of six-membered rings and are accordingly not subject to large-amplitude

[172] P. J. Buckley, C. C. Costain, and J. E. Parkin, *Chem. Commun.* 668 (1968); *Symp. Gas Phase Mol. Struct., 2nd, Austin, Texas* Paper M-12 (1968).

[173] J. J. Sloan and R. Kewley, *Can. J. Chem.* **47**, 3453 (1969); **50**, 955 (1972).

[174] S. C. Dass and R. Kewley, *Symp. Mol. Struct. Spectrosc., Columbus, Ohio*, Paper P9 (1973).

[175] E. Hirota and S. Suenaga, *J. Mol. Spectrosc.* **42**, 127 (1972).

[176] E. Hirota, *J. Mol. Spectrosc.* **38**, 367 (1971).

internal motions. The spectrum of 2,8,9-trioxaadamantane[177] is that of a symmetric rotor with a dipole moment of 3.01 D.

VIII. FUTURE DEVELOPMENTS

Although many major contributions to our knowledge of heterocyclic structures have now been made by the microwave method, it is clear that, with the techniques as at present established, much work in the immediate future will considerably enlarge the range of fundamental ring structures for which details are known with accuracy. For the larger cyclic molecules, perhaps the chief interest will be in establishing conformations and their relative energies, and in combined applications of spectroscopy and electron diffraction.[56]

As the powers of the methods are increased, we may expect improved accuracy of detail. Beam maser spectroscopy, with its greatly increased resolution, is already being applied to simple heterocyclic systems,[178] yielding hyperfine coupling constants for magnetic interactions of the protons in furan[179] and similar information. With current improvements in sensitivity of such methods, they will become applicable to many of the more simple heterocyclic systems and, for example, greatly improve the precision of our knowledge of nitrogen quadrupole coupling constants.

ACKNOWLEDGMENT

The author is grateful to a number of colleagues who have provided unpublished details of recent work.

[177] W. D. Slafer and D. O. Harris, *J. Mol. Spectrosc.* **45**, 412 (1973).
[178] G. R. Tomasevitch and K. D. Tucker, *Symp. Mol. Struct. Spectrosc., Columbus, Ohio* Paper Z7 (1972); private communication.
[179] G. R. Tomasevitch, K. D. Tucker, and P. Thaddeus, *J. Chem. Phys.* **59**, 131 (1973).
[180] L. N. Gunderova, A. A. Shapkin, and N. M. Pozdeev, *Opt. Spektrosk.* **34**, 1211 (1973).
[181] S. Doraiswamy and S. D. Sharma, private communication.
[182] J. Mjöberg and S. Ljunggren, *Z. Naturforsch.* **A28**, 729 (1973).

· 3 ·

ESR Spectroscopy of Heterocyclic Radicals

B. C. GILBERT and M. TRENWITH

DEPARTMENT OF CHEMISTRY, UNIVERSITY OF YORK

HESLINGTON, YORK, YO1 5DD, ENGLAND

In recent years, applications of electron spin resonance (ESR) spectroscopy to the investigation of organic radicals have become increasingly numerous and sophisticated. The particular attractions of this technique include its high sensitivity (a radical concentration of 0.1 μM in solution is readily detectable) and its ability to yield detailed information concerning many radical properties; a full analysis of an ESR spectrum can lead not only to the recognition that a certain radical participates in a chemical reaction but also to structural and conformational details of the radical concerned, together with kinetic and thermodynamic parameters for a range of intra- and intermolecular processes.

This chapter is devoted to a discussion of information which has been obtained from the *isotropic* spectra (i.e., for fluid solutions) of free radicals from heterocyclic compounds. An introductory description of the parameters used to characterize ESR spectra is followed by a review of the contribution of ESR studies to our understanding of the radical reactions exhibited by alicyclic compounds containing heteroatoms. This includes reference to the hybridization, conformation, and further reactions of the heteroalicyclic radicals involved. Aromatic radicals of the cation, anion, and neutral varieties are then discussed, with the emphasis placed mainly on their electronic structure. Finally, brief consideration will be given to a miscellany of radicals, including some of biological importance. Methods of radical generation are described where appropriate.

I. ESR PARAMETERS

The following brief summary should serve to establish the usefulness of the three essential parameters which characterize an ESR spectrum: g values, hyperfine splittings, and linewidths. Detailed theory can be found in the books by Carrington and McLachlan,[1] Ayscough,[2] and Wertz and Bolton,[3] while that by Alger[4] deals with instrumentation. Reviews which provide further detail in certain areas include those on organic radicals in solution and solids,[5] short-lived radicals in solution,[6] and radical conformations.[7] Ion radicals, especially those from aromatic and heterocyclic species, have been discussed at some length in both a review[8] and a monograph.[9] A book devoted to biological applications of ESR has also recently appeared.[10]

[1] A. Carrington and A. D. McLachlan, "Introduction to Magnetic Resonance," Harper, New York, 1967.

[2] P. B. Ayscough, "Electron Spin Resonance in Chemistry," Methuen, London, 1967.

[3] J. E. Wertz and J. R. Bolton, "Electron Spin Resonance: Elementary Theory and Practical Applications." McGraw-Hill, New York, 1972.

[4] R. S. Alger, "Electron Paramagnetic Resonance: Techniques and Applications." Wiley (Interscience), New York, 1968.

[5] M. C. R. Symons, *Advan. Phys. Org. Chem.* **1**, 283 (1963).

[6] R. O. C. Norman and B. C. Gilbert, *Advan. Phys. Org. Chem.* **5**, 53 (1967).

[7] D. H. Geske, *Progr. Phys. Org. Chem.* **4**, 125 (1967).

[8] A. Carrington, *Quart. Rev., Chem. Soc.* **17**, 67 (1963).

[9] F. Gerson, "High Resolution E.S.R. Spectroscopy," Vol. 1, Chemical Topics for Students. Wiley (Interscience), New York, 1970.

[10] H. M. Swartz, J. R. Bolton, and D. C. Borg (eds.), "Biological Applications of Electron Spin Resonance." Wiley (Interscience), New York, 1972.

A. g Values

If a solution containing free radicals (and hence unpaired electrons) is placed within the cavity of an ESR spectrometer operating at a fixed frequency ν, then the basic equation (1) indicates the magnitude of the

$$h\nu = g\beta H_0 \qquad (1)$$

magnetic field which must be applied for resonance absorption to take place. Both h, Planck's constant, and β, the Bohr magneton, are fundamental constants so that the factor g is characteristic of the paramagnetic species under investigation. For the free electron, $g_e = 2.0023$; for microwave radiation of fixed frequency about 9500 MHz (Radar X-band) absorption should take place when the variable field is adjusted so that the resonance equation is fulfilled (about 3400 gauss or Oe, these units being used interchangeably; the SI equivalent of the oersted is the tesla, $1T = 10^4$ Oe).*

For organic radicals, contributions to the g value from the orbital angular momentum of the unpaired electron are almost completely quenched and the g values are close to 2.0023. However, small g shifts (and hence differences in the magnetic field required for resonance) do arise when spin–orbit interactions mix σ and π configurationally excited states into the ground state. The magnitude of the interaction depends on the electron spin density distribution, being most marked for radicals in which considerable spin density resides on heavy atoms (the spin–orbit coupling constant depends on the fourth power of the atomic number) and where low-lying excited electronic states are available. Thus spin–orbit interactions are less significant for $\cdot CH_3$ ($g = 2.0026$) than for $\cdot CH_2OH$ ($g = 2.0033$), where spin is delocalized on to oxygen

$$\dot{C}H_2\text{—}\ddot{O}H \longleftrightarrow \overset{-}{\ddot{C}}H_2\text{—}\overset{+}{\ddot{O}}H$$

and where $n \rightarrow \pi^*$ transitions involving promotion of a nonbonding electron into the half-filled orbital are of low energy. The difference in g for these radicals corresponds to approximately 1.3 G, which is 0.13 mT (millitesla), on an X-band spectrometer.

Further examples are provided by the series of cation radicals (1, X = O, S, Se) where the g values are 2.0032, 2.0053, and 2.0161,

---

* The reader is referred elsewhere for discussions of the dynamic relaxation processes involved and for consideration of the requisite instrumentation. The special problems associated with triplets and biradicals are beyond the scope of this review.

respectively.[11] Similar trends—with characteristically different magnitudes—are observed for the corresponding nitroxide $\left(\diagdown\text{N—O}\cdot\right)$ and neutral $\left(\diagdown\text{N}\cdot\right)$ radical series. g Values for organic radicals are

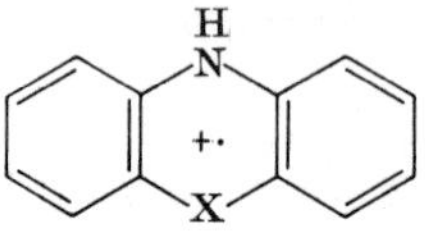

H
N

+·

X

[1]

clearly a potential source of much important information and can be particularly useful in studies of heterocyclic species as sensitive probes for the extent of odd-electron delocalization on to intracyclic heteroatoms. It should be evident that the g value is also a useful diagnostic property (cf. NMR τ values). Interestingly, many σ radicals (in which the unpaired electron resides in an orbital with considerable s character, e.g., $\cdot$CHO) have g values lower than that of the free electron.

B. Hyperfine Splittings

Hyperfine splittings arise from an interaction between the unpaired electron and the magnetic moments of appropriate nuclei within the same radical. Those nuclei from which splittings are often observed include ^{1}H, ^{19}F, and ^{31}P (which have nuclear spin $I = \frac{1}{2}$), ^{14}N ($I = 1$), and ^{35}Cl, ^{37}Cl, ^{79}Br, and ^{81}Br ($I = \frac{3}{2}$). Enrichment techniques are generally necessary for the observation of splittings from low-abundance isotopes such as ^{13}C ($I = \frac{1}{2}$, 1.11% abundance), ^{33}S ($I = \frac{3}{2}$, 0.76%), and ^{17}O ($I = \frac{5}{2}$, 0.037%). The appearance of ESR spectra from radicals containing one or more interacting nuclei (e.g., a splitting into a doublet, 1:1, produced by a single nucleus with spin $\frac{1}{2}$) has been adequately exemplified, [1-10] so attention here will be devoted to a short survey of information about radicals obtainable from their hyperfine splittings.

Isotropic splittings, exhibited by radicals tumbling freely in solution,* derive from a direct interaction (Fermi contact) between the unpaired electron and the nuclei of magnetic isotopes in the same species. They should arise, therefore, only when the square of the wavefunction describing the unpaired electron has a finite value at the nucleus concerned; $|\psi(0)|^2 \neq 0$.

That radicals of p or π type, such as the planar radical $\cdot$CH$_3$ (in

---

* The anisotropic dipole–dipole interaction is then averaged to zero.

[11] M. F. Chiu, B. C. Gilbert, and P. Hanson, *J. Chem. Soc. B* 1700 (1970).

which it is expected that $|\psi(0)|^2 = 0$ because the nuclei lie in the nodal plane of the p orbital) do show hyperfine splitting is due to the mechanism of *spin polarization* which is represented schematically in Fig. 1.

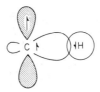

FIG. 1. Schematic representation of spin polarization for a ·C—H fragment.

The unpaired electron in the p_z orbital on carbon effects a slight decoupling of the paired electrons in the σ molecular orbital adjacent to it, in the sense indicated. Thus the carbon nucleus experiences spin of the same sign as that in the p_z orbital while that produced in the hydrogen $1s$ orbital is of opposite sign. The hyperfine splittings for this radical[12] are $a(^{13}C) = +3.85$ mT (38.5 G) and $a(H) = -2.3$ mT (-23 G); signs of ESR splittings, however, are not usually determined. The ESR spectrum of $^{12}CH_3$ consists of a $1:3:3:1$ pattern, the separation between adjacent lines being 2.3 mT.

For the methyl radical, the "spin density" in the p_z orbital is unity; in cases where the spin density, ρ_π, in the p_z orbital of a $\overset{\displaystyle\cdot}{\diagdown}$C—H fragment is less than unity, $a(H)$ is proportionately reduced:

$$a(H) = Q^H \rho_\pi \qquad (2)$$

In practice, the McConnell[13] equation (2) is fairly well obeyed with $Q^H = (-)2.3$ mT [cf. $a(H)$ for ·CH$_3$] and it is widely used for estimating spin densities (ρ_π) at radical centers from the attached hydrogen's splitting (these are referred to as α-hydrogen splittings). For example, in the ESR spectrum of the dibenzoselenophene anion radical [2], which is shown in Fig. 2, the two $1:2:1$ splittings of 0.518 and 0.421 mT from the 2, 7, and 4, 5 hydrogens, respectively, indicate that approximately $0.518/2.3$ and $0.421/2.3$, that is 0.225 and 0.183, respectively, of the unpaired electron density resides in the corresponding $2p_z$ orbitals. The $1:4:6:4:1$ splitting of 0.103 mT from the remaining hydrogens indicates that the fraction is $0.103/2.3$, i.e., 0.045, for the carbon atoms at

---

[12] R. W. Fessenden and R. H. Schuler, *J. Chem. Phys.* **43**, 2704 (1965).

[13] H. M. McConnell, *J. Chem. Phys.* **24**, 764 (1956); H. M. McConnell and D. B. Chesnut, *ibid.* **28**, 107 (1958).

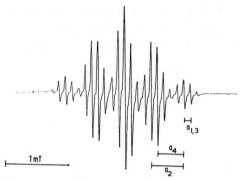

Fig. 2. ESR spectrum of the anion radical of dibenzoselenophene [2].

positions 1, 3, 6, and 8. Delocalization of the unpaired electron around the π perimeter in the nitroxide [3] derived from phenoselenazine leads to splittings from two sets of four apparently equivalent hydrogens

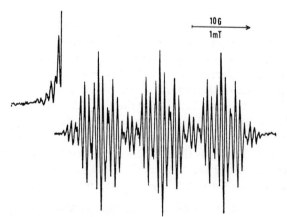

[2]　　　　　[3]

[$a(H)$ = 0.21 mT for hydrogens at positions 1, 3, 7, and 9, and $a(H)$ = 0.06 mT for the remainder; the assignment is made on the basis of simple VB considerations and MO calculations]. The spectrum (Fig. 3)

Fig. 3. ESR spectrum of phenoselenazine nitroxide [3] with (inset) the low-field satellites from ^{77}Se, recorded at high gain.

also shows a splitting (1:1:1) from nitrogen, resulting from unpaired electron density on this atom.

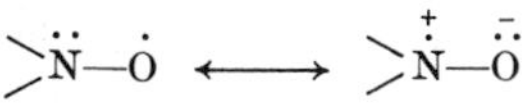

$$\ce{>\overset{..}{N}-\overset{.}{O} <---> >\overset{+}{N}-\overset{-}{\underset{..}{O}}}$$

Satellite lines from those species which contain ^{77}Se($I = \frac{1}{2}$, 7.58%) can also be detected for radicals *2* and *3* (see Fig. 3). The relationships between splittings for nuclei other than ^{1}H and the spin density distribution will be discussed later.

Hyperfine splittings are also observed for β hydrogens in π radicals (e.g., $\ce{>\overset{.}{C}-CH_3}$), and these are thought to arise via hyperconjugation:

$$\ce{.CH_2-CH_2^{H} <---> CH_2=CH_2^{.H}}$$

For the ethyl radical, $a(\alpha\,H) = 2.238$ and $a(\beta\,H) = 2.687$ mT.[14] The magnitude of a β-hydrogen splitting depends on the spin density on the adjacent (α) carbon atom; the proportionality constant Q^{CH_3} which links ρ_α to $a(CH_3)$ is 2.93 mT, and this can be used to derive spin densities at radical centers from splittings for freely rotating methyl groups. The β-hydrogen splitting also depends, at any given instant, on the dihedral angle, θ, between the β C—H bond and the half-filled orbital. In practice, a $\cos^2\theta$ relationship seems to be obeyed, and it is manifest, for example, for cyclic radicals in which the conformation is locked or in which inter-conversion is "slow" (see Section I, C); for the radical *4* (X = Me), splittings are 2.63 and 0.385 mT for the axial and equatorial β hydrogens, respectively, at $-100°C$.[15] Applications of ESR to the conformational analysis of nitroxides, including cyclic examples related to that

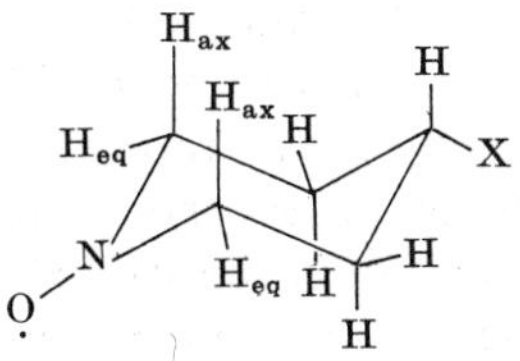

[*4*]

depicted here, have been reviewed.[16] The angular dependence of a β-hydrogen splitting leads to values for $a(\beta\,H)$ in radicals $\ce{>\overset{.}{C}-CH_2X}$

[14] R. W. Fessenden and R. H. Schuler, *J. Chem. Phys.* **39**, 2147 (1963).
[15] J. J. Windle, J. A. Kuhnle, and B. H. Beck, *J. Chem. Phys.* **50**, 2630 (1969).
[16] E. G. Janzen, *Top. Stereochem.* **6**, 177 (1971).

and $\diagdown\overset{\centerdot}{C}$—$CHX_2$ which may be different from those for the analogous $\diagdown\overset{\centerdot}{C}$—$CH_3$ radicals, and which indicate the nature of the preferred conformation of the substituent(s); detailed examples can be found in the review by Geske.[7]

Hyperfine splittings from γ and δ hydrogens in π radicals are not normally observed, with the notable exception of some interesting long-range interactions which reflect certain rigid, favorable pathways for transmission of spin density in some cyclic systems. In σ radicals, such as $\cdot CHO$ and $\cdot CF_3$, large isotropic interactions may occur since the orbital containing the unpaired electron has s character and the nuclei concerned are not in its nodal plane.

C. Linewidths

The linewidth (ΔH) characteristic of a spectrum—the distance, in field units, between the maximum and minimum for a resolved peak in the first derivative presentation—is usually small for organic radicals in solution (0.001–0.1 mT). In most cases, ΔH includes a contribution from anisotropy in both the g value and the hyperfine splitting and increases with increasing solvent viscosity and decreasing temperature. Other interactions (usually unwanted) which can cause line broadening by shortening relaxation times include spin rotation and, sometimes, quadrupole coupling.

However, rate processes of a chemical nature also exist which can shorten the lifetime of a spin state and lead to an uncertainty in the electronic energy levels, which of course gives rise to line broadening; sometimes this can yield useful chemical information. For example, if a radical undergoes rapid electron exchange with a neutral molecule then the consequent broadening can be used to obtain rate data for this exchange; instances involving ion radicals from heterocycles undergoing one-electron exchange with their parent molecules have been reported.[17] Line broadening of a slightly different kind—the so-called alternating linewidth effect—results when interconversion of conformations, for example, takes place at a rate between those limits characteristic of "slow" and "fast" exchange. For the piperidine nitroxide (4, $X = H$) the spectrum at $-103°C$ shows coupling to two pairs of β hydrogens ("slow" exchange) whereas at $110°C$ the four hydrogens couple equivalently ("fast" exchange). At intermediate

[17] S. P. Sorensen and W. H. Bruning, *J. Amer. Chem. Soc.* **94**, 6352 (1972); W. L. Reynolds, *J. Phys. Chem.* **67**, 2866 (1963).

temperatures—where the rate of interconversion of equivalent chair conformations is approximately the same as the difference between the splittings of hydrogens whose positions are being exchanged, i.e., about 2.3 mT or 64 MHz in this case—broadening results. Analysis of the linewidths leads to more exact rate data; for this example the activation enthalpy for ring inversion was calculated as 22.6 kJ mole^{-1}.[15]

Finally, the usefulness of ESR as a method for obtaining kinetic data by monitoring rates of radical decay should not be overlooked. The scope of the method is illustrated by recent determinations[18] of the rate constants for reaction of hydrogen atoms (generated in solution by steady state radiolysis) with a variety of aromatic and heterocyclic compounds. Rate constants in the range 0.6–3.3 × 10^8 liters mole^{-1} sec^{-1} were reported.

II. HETEROALICYCLIC RADICALS

A. Radicals from Cyclic Ethers, Thioethers, Amines, and Related Compounds

The application of ESR to the direct detection of transient radical intermediates stems from the development of flow methods and photolytic techniques for the generation of sufficiently high steady state radical concentrations. The former method involves, typically,[19] the mixing of aqueous solutions containing Ti(III), H_2O_2 (to give ·OH), and a substrate, in a rapid-flow system. Radicals derived by reaction of the hydroxyl radical with the substrate [e.g., tetrahydrofuran (THF), piperidine, 1,4-dioxane[19]] can then be detected. In the second method, solutions of the substrate and, for example, di-t-butyl peroxide or hydrogen peroxide (sometimes in an inert diluent) are irradiated *in situ* using a suitable source (usually a high pressure mercury lamp); reference 20 describes a typical application of this technique to the investigation of the reactions of ·O-t-Bu with some cyclic ethers. The scope of both methods has been extended to the preparation of other reactive species including ·SR, ·NR$_2$, and ·SO$_4$$^-$.

For a variety of cyclic ethers and thioethers,[19–22] reaction with ·OH [Ti(III)–H_2O_2] or ·O-t-Bu leads to the appropriate conjugated radicals, the site of hydrogen atom abstraction apparently being

[18] P. Neta and R. H. Schuler, *J. Amer. Chem. Soc.* **94**, 1056 (1972); P. Neta, *Chem. Rev.* **72**, 533 (1972).

[19] W. T. Dixon and R. O. C. Norman, *J. Chem. Soc.* 4850, (1964).

[20] A. Hudson and K. D. J. Root, *Tetrahedron* **25**, 5311 (1969).

[21] I. Biddles, A. Hudson, and J. T. Wiffen, *Tetrahedron* **28**, 867 (1972).

[22] E. A. C. Lucken and B. Poncioni, *Helv. Chim. Acta* **55**, 2673 (1972).

determined by the electrophilic nature of the attacking species.[6] Thus tetrahydrofuran reacts with the Ti(III)–H_2O_2 system[19] to give a predominance of the signal from the α radical [5] over that from the β radical [6] (splittings given here and elsewhere in this review are in mT); however, the reverse has been observed for the reaction with Fe(II)–H_2O_2.[23] This behavior has been rationalized,[24] in terms not of there

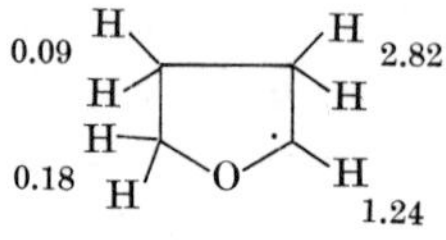

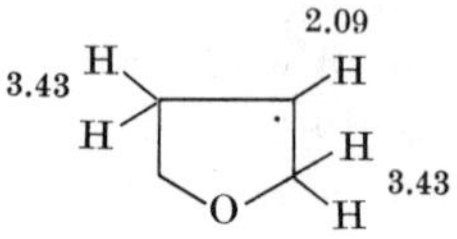

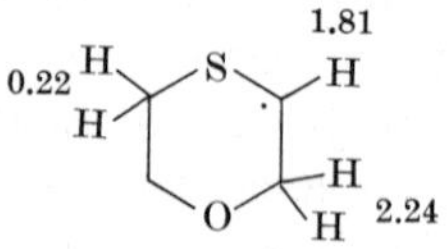

being different reactive intermediates in the two cases but rather of the ready oxidation of the oxygen-conjugated radical 5 to the corresponding carbonium ion by Fe(III) but not by Ti(IV). In both cases it seems that radical 5 is formed faster than radical 6. The formation of 7 rather than the oxygen-conjugated radical, in the reaction of ·O-t-Bu with the parent heterocycle,[21] may be due at least in part to relatively facile electron abstraction at sulfur followed by deprotonation.[25]

Hyperfine splittings for these and related radicals have been interpreted in terms both of the conformational properties of the radicals and of the relative delocalizing abilities of oxygen and sulfur. It emerges that, as judged by α and β hydrogen splittings,[24–26] a sulfur atom conjugated to a radical center

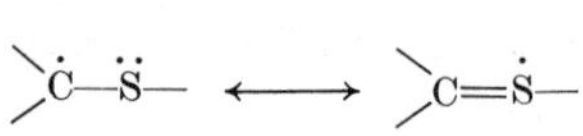

$$\overset{\cdot}{\underset{/}{>}}C\!-\!\overset{\cdot\cdot}{S}\!-\ \longleftrightarrow\ \underset{/}{>}C\!=\!\overset{\cdot}{S}\!-$$

withdraws more unpaired electron density than oxygen (about 25% compared with 15%), but that in the latter case considerable bending at the radical center also results. This has been attributed to the $+M$ effect of oxygen,

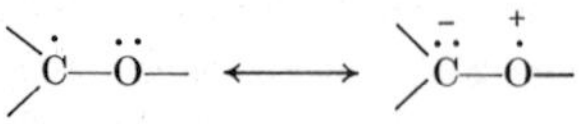

$$\overset{\cdot}{\underset{/}{>}}C\!-\!\overset{\cdot\cdot}{\underset{\cdot\cdot}{O}}\!-\ \longleftrightarrow\ \underset{/}{>}\overset{\bar{}}{C}\!-\!\overset{+}{O}\!-$$

which renders the radical center carbanion-like,[26] and the resulting bending causes anomalously low splittings and sometimes even *positive* values for $a(\alpha\ H)$. Thus although the effect is not very marked in 8, the

[23] T. Shiga, A. Boukhors, and P. Douzou, *J. Phys. Chem.* **71**, 4264 (1967).
[24] R. O. C. Norman and P. R. West, *J. Chem. Soc. B* 389 (1969).
[25] B. C. Gilbert, J. P. Larkin, and R. O. C. Norman, *J. Chem. Soc. Perkin Trans. 2*, 272 (1973).
[26] A. J. Dobbs, B. C. Gilbert, and R. O. C. Norman, *J. Chem. Soc. A* 124 (1971).

very small α-hydrogen splitting (mT) in *9* and the (assumed) positive values for *10* and *11* probably indicate markedly increased bending. The increased effects for *10* and *11* over *9* probably indicate, respectively,

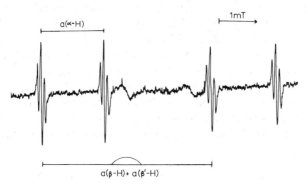

the better lone-pair conjugation [*10*] and the strain in the ring [*11*] compared to the six-membered heterocycle.

For the oxiranyl radical [*11*] the β hydrogens have an anomalously low splitting (probably a consequence of the radical bending) and appear equivalent; evidently inversion at the radical center is rapid under the conditions of observation (> 10^8 sec^{-1} at room temperature). From the alternating line broadening for the abstraction radical from 1,4-dioxane [*12*] at about room temperature (Fig. 4), a rate of ring inversion of the

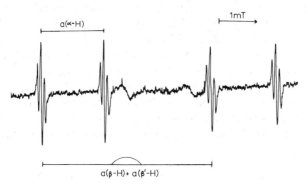

FIG. 4. ESR spectrum of the 1,4-dioxanyl radical [*12*] showing alternating linewidths.

order of 5 × 10^8 sec^{-1} has been estimated.[19] Interestingly, although the splittings for *12* and related ether-type radicals can be accommodated

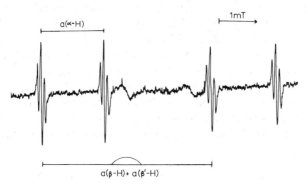

105

in terms of a chairlike radical, in *13* the low β-hydrogen splitting leads to the suggestion that the radical prefers a conformation with the β-sulfur atom eclipsing the orbital of the unpaired electron.[21] Sulfur d orbital participation for the β substituent may be implied.

Electron spin resonance studies of a wide range of alicyclic oxygen-containing radicals, generated through reactions of hydroxyl and t-butoxyl, have shown no evidence of further radicals formed by ring opening. Thus the oxiranyl radical appears to remain intact,[26,27] and radical *14*, obtained from the parent 1,3-dioxolane, is apparently not converted into either *15* or *16* nor is it formed from them. However, the conversion of *15* into *16* has been demonstrated.[27]

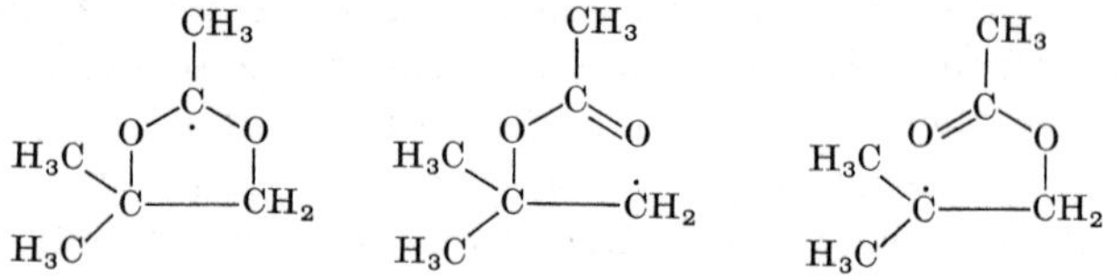

[14]   [15]   [16]

In another investigation,[28] photochemical excitation of benzophenone was employed to abstract hydrogen atoms from 2-alkyl-1,3-dioxolanes (*17*, X = O) and 2-alkyl-1,3-oxathiolanes (*17*, X = S) in benzene, and the resulting radicals [*18*] were trapped with 2-methyl-2-nitrosopropane to give the relatively stable nitroxides [*19*]; there was no evidence for ring opening even when R = cyclopropyl (see Janzen[29] for a review of spin trapping techniques). When the photolysis is carried out with $CFCl_3$ as solvent, ring-opened products are derived by heterolysis of chlorides produced via radical abstraction reactions. Interestingly,

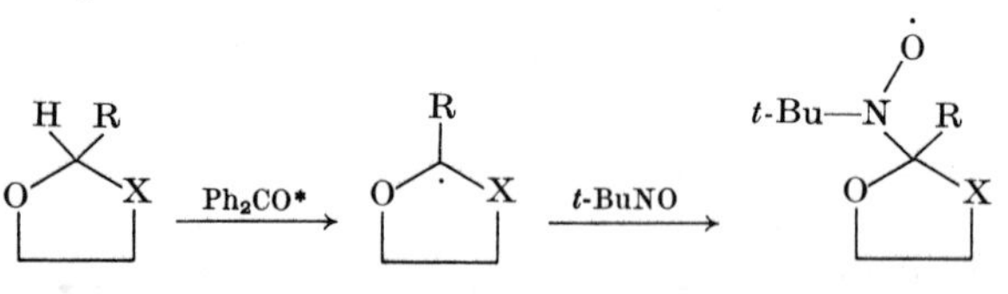

[17]   [18]   [19]

reaction of propylene oxide with the hydroxyl radical [Ti(III)–H_2O_2] in a flow system leads to the formation of isomeric hydroxyallyl radicals,[30] probably via the reaction shown in Eq. (3). Another interesting

[27] A. L. J. Beckwith and P. K. Tindal, *Aust. J. Chem.* **24**, 2099 (1971).

[28] J. W. Hartgerink, L. C. J. van der Laan, J. B. F. N. Engberts, and Th. J. de Boer, *Tetrahedron* **27**, 4323 (1971).

[29] E. G. Janzen, *Accounts Chem. Res.* **4**, 31 (1971).

[30] A. J. Dobbs, B. C. Gilbert, and R. O. C. Norman, unpublished observations.

radical transformation, involving the acid-catalyzed elimination of water, has been reported for suitably substituted radicals derived by oxidation of some monosaccharides with the hydroxyl radical[31] [Eq. (4)].

$$
\begin{array}{ccc}
\underset{\substack{|\\ \text{CH} \\ \diagup\diagdown \\ H_2C\!\!-\!\!O}}{CH_3} & \overset{\cdot OH}{\longrightarrow} & \underset{\substack{|\\ \text{CH} \\ \diagup\diagdown \\ H_2C\!\!-\!\!O}}{\cdot CH_2} & \longrightarrow & \underset{\substack{\vdots\\ \text{CH} \\ \diagup\diagup \\ HC\!\!-\!\!OH}}{\vdots CH_2}
\end{array} \tag{3}
$$

$$
\begin{array}{ccc}
\overbrace{\underset{CH(OH)\!-\!CH(OH)}{O}} & \overset{\cdot OH}{\longrightarrow} & \overbrace{\underset{\dot C(OH)\!-\!CH(OH)}{O}} & \overset{-H_2O}{\longrightarrow} & \overbrace{\underset{\underset{\displaystyle O}{\overset{\|}{C}}\!-\!\dot CH}{O}}
\end{array} \tag{4}
$$

Other cyclic radicals generated using the Ti(III)–H_2O_2 reaction include those from ethylene carbonate, propylene carbonate, 2-oxazolidone, and some related compounds,[32] and also some 7-oxabicyclo-[2.2.1]hept-2-yl radicals.[33]

Electron spin resonance results for the reactions of heteroalicyclic compounds containing nitrogen are less clearly defined. It has been pointed out[19] that under the acidic conditions used for most Ti(III)–H_2O_2 studies amino groups will be protonated. This probably accounts for the observation that with piperidine abstraction of a C(3) hydrogen atom takes place (since ·OH is electrophilic) and that an oxygen-conjugated radical is produced from morpholine. These observations cast doubt on the assignment of the spectrum from pyrrolidine,[34] obtained under acidic conditions, to the nitrogen-conjugated radical.

The aziridinyl [20] and azetidinyl [21] radicals have been generated[35] by reaction of the parent amines with t-butoxyl radicals (generated photolytically) in cyclopropane at − 100°C; the coupling constants (mT) and g values suggest, in agreement with INDO (Intermediate Neglect of Differential Overlap) molecular orbital calculations, that the unpaired electron resides in a $2p_z$ orbital on nitrogen. Under these photolytic conditions no evidence for hydrogen migration in the aminyl

[31] R. O. C. Norman and R. J. Pritchett, J. Chem. Soc. B 1329 (1967); see also P. J. Baugh, O. Hinojosa, and J. C. Arthur, J. Phys. Chem. 71, 1135 (1967).

[32] C. Corvaja, M. Brustolon, and G. Giacometti, Z. Phys. Chem. (Frankfurt) 66, 279 (1969).

[33] T. Kawamura, T. Koyama, and T. Yonezawa, J. Amer. Chem. Soc. 92, 7222 (1970).

[34] H. Taniguchi, J. Phys. Chem. 74, 3143 (1970).

[35] W. C. Danen and T. T. Kensler, Tetrahedron Lett. 2247 (1971).

$$
\begin{array}{cc}
[20] & [21] \\
g = 2.0043 & g = 2.0045 \\
1.252 & 1.399
\end{array}
$$

radicals was observed,[36] though acid-catalyzed conversion of alkyl-aminyl radicals into α-aminoalkyl radicals has been demonstrated in another context.[37] Isotropic spectra of the pyrrolidinyl and Δ^3-pyrro-linyl radicals, generated in an adamantane matrix at 77°K by X irradi-ation of the parent amines, have also been recorded.[38]

B. Cyclic Nitroxide Radicals

Oxidation of amines in the presence of peroxides or oxygen has often led to the production of nitroxide radicals ($R_2\ddot{N}$—$\dot{O}$ ↔ $R_2\overset{+}{N}$—$\overset{-}{O}$) which have not always been recognized as such or even wanted. The number of ESR studies concerned with nitroxides probably outweighs that of any other single category of organic radical, and detailed attention has been paid to their synthesis, reactivity, structure, and mediation in radical processes. Some nitroxides, especially certain cyclic varieties, are extremely stable, and this has led to their application in the "spin-probe" or "spin-labeling" technique (e.g., for biomolecules[39]); their involvement in spin-trapping experiments will already be apparent.

Aziridine nitroxide [22] is as yet unknown. Attempts in the authors' laboratory to prepare it by the oxidation of ethyleneimine using basic, aqueous hydrogen peroxide with phosphotungstic acid as a catalyst gave a signal evidently derived from ring-opened polymeric material with randomly oxidized amine functions. A recent claim[40] for production of the tetramethyl derivative of 22 has been refuted.[41] In contrast, a few four-membered ring nitroxides are known; for example,[42] 2,2,4,4-

---

[36] W. C. Danen and T. T. Kensler, *J. Amer. Chem. Soc.* **92**, 5235 (1970).

[37] N. H. Anderson and R. O. C. Norman, *J. Chem. Soc. B* 993 (1971).

[38] D. W. Pratt, J. J. Dillon, R. V. Lloyd, and D. E. Wood, *J. Phys. Chem.* **75**, 3486 (1971).

[39] O. H. Griffith and A. S. Waggoner, *Accounts Chem. Res.* **2**, 17 (1969); I. C. P. Smith, *in* "Biological Applications of Electron Spin Resonance" (H. M. Swartz, J. R. Bolton, and D. C. Borg eds.), Chapter 11. Wiley (Interscience), New York, 1972.

[40] G. R. Luckhurst and F. Sundholm, *Tetrahedron Lett.* 675 (1971).

[41] P. Singh, D. G. B. Boocock, and E. F. Ullman, *Tetrahedron Lett.* 3935 (1971); J. F. W Keana, R. J. Dinerstein, and D. P. Dolata, *ibid.* 119 (1972).

[42] J.-C. Espie and A. Rassat, *Bull. Soc. Chim. Fr.* 4385 (1971).

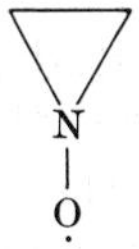

H OH

Me   Me

Me   Me

N   N

| |

O   O

[22] 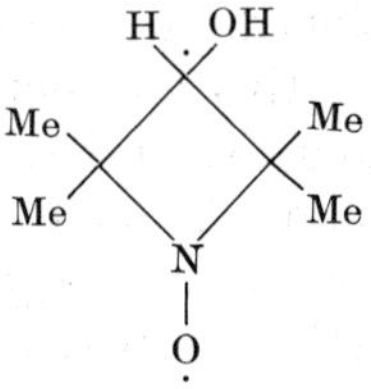   [23]

tetramethyl-3-hydroxyazetidine nitroxide [23], produced by treatment of the parent amine with hydrogen peroxide and phosphotungstic acid, is quite stable (e.g., in aqueous solution at room temperature); this leads to the suggestion that it might be useful as a spin label.

The many detailed investigations of 5- and higher-membered alicyclic nitroxides are too numerous for a complete coverage here. Rather, we draw attention to two reviews concerned mainly with stereochemistry[16,43] and we attempt to highlight some of the more important conclusions.

Spectra have been recorded as a function of temperature for a variety of unhindered piperidinyl nitroxides, prepared from the corresponding hydroxylamines by photolysis of solutions in dichloromethane,[15,44] and also for a variety of other cyclic nitroxides (including those from morpholine,[45] piperazine,[46] and hepta- and octamethyleneimine[47]) prepared in static, aqueous systems with hydrogen peroxide.

The ESR spectrum of pyrrolidine nitroxide [24] shows a splitting of 2.23 mT from four apparently equivalent β hydrogens. No dependence of the spectrum on temperature is evident, so it may be concluded that the magnetic environments of the β hydrogens are completely averaged by a process, probably pseudorotation, which has a small activation energy and which is therefore rapid even at low temperatures.[45] The behavior of piperidine nitroxide (25, R = H) at a series of temperatures

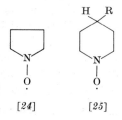

[24]   [25]

[43] A. Rassat, *Pure Appl. Chem.* **25**, 623 (1971).
[44] R. E. Rolfe, K. D. Sales, and J. H. P. Utley, *J. Chem. Soc. D* 540 (1970).
[45] A. Hudson and H. A. Hussain, *J. Chem. Soc. B* 251 (1968).

has already been mentioned, and various studies[15,44,45] have produced good agreement for the activation energy for ring inversion, as measured from the line broadening. Several radicals of the type *25*, with R = Me, *i*-Pr, *n*-Pr, *t*-Bu, and Ph, have been studied closely, and strong conformational preferences are apparent.[15,44,46] For example, at 80°C, 90% of the 4-methyl derivative has the substituent equatorial; for the 4-phenyl derivative at 140°C the proportion is 96%.[15] At room temperature the conformational preferences of the 4-methyl and 4-isopropyl groups are 7.1 and 8.8 kJ mole^{-1}, respectively.[44] Fluxional motion in larger rings has been discussed.[47]

Radicals with the nitroxide function flanked by tertiary carbon atoms are often considerably more stable than those discussed above and continuous generation is not usually necessary. For bicyclic examples the ESR spectra often reveal instances of long-range interaction with suitably sited γ, and sometimes δ, hydrogens, and detailed analysis in some of these cases has considerably aided our understanding of the stereochemical requirements for such interaction. For example in *26*, which was prepared by oxidation of the corresponding bicyclic

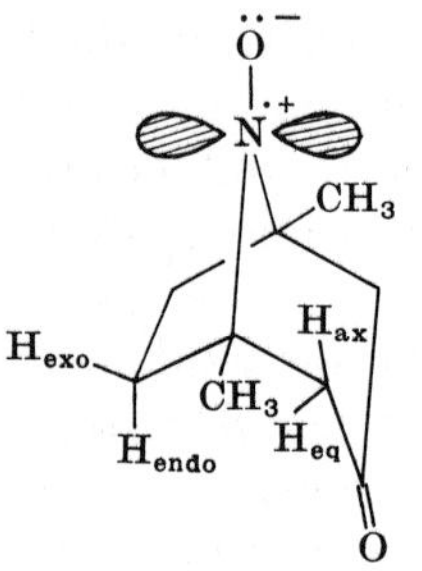

$a(N)$	1.95 mT
$a(H_{ax})$	-0.11
$a(H_{eq})$	$+0.25$
$a(H_{endo})$	-0.015
$a(H_{exo})$	-0.030
$a(CH_3)$	-0.044

[*26*]

[3.2.1]amine with H_2O_2-phosphotungstic acid,[48] the ESR spectrum was analyzed in conjunction with NMR (the latter method, which is applicable to some stable radicals, gives the signs, as well as the magnitudes, of the hydrogen splittings; see also Hatch and Kreilick[49]). Comparison of the observed splittings was made with those calculated by various theoretical procedures. One method incorporating only through-bond spin polarization gives a good account of the negative γ splittings but not the positive, stereoselective $a(H_{eq})$ associated with the "W-plan"

[46] A. Hudson and H. A. Hussain, *J. Chem. Soc. B* 953 (1968).
[47] A. Hudson and H. A. Hussain, *J. Chem. Soc. B* 1346 (1968).
[48] A. Rassat and J. Ronzaud, *J. Amer. Chem. Soc.* 93, 5041 (1971).
[49] G. F. Hatch and R. W. Kreilick, *J. Chem. Phys.* 57, 3696 (1972).

interaction (homohyperconjugation; see *27*). In contrast, INDO calculations give good agreement.

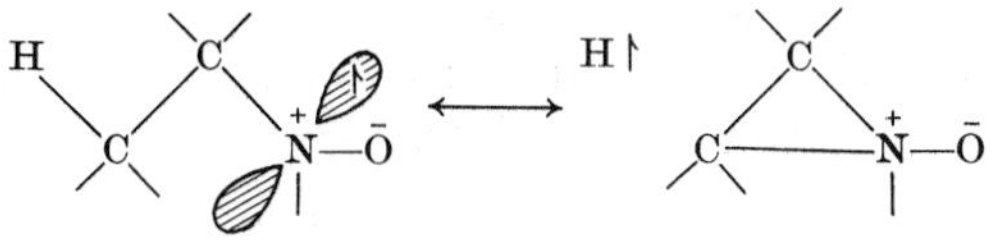

[*27*]

Some stable isoquinuclidine nitroxides have been prepared,[50,51] and in one case [Eq. (5)] a 72% yield (of *28*) was obtained.[51] Reasonable yields have also been reported for syntheses of the stable pyrrolidinyl nitroxides *29* and *30* by reactions shown in Eqs. (6) and (7), respectively.[52]

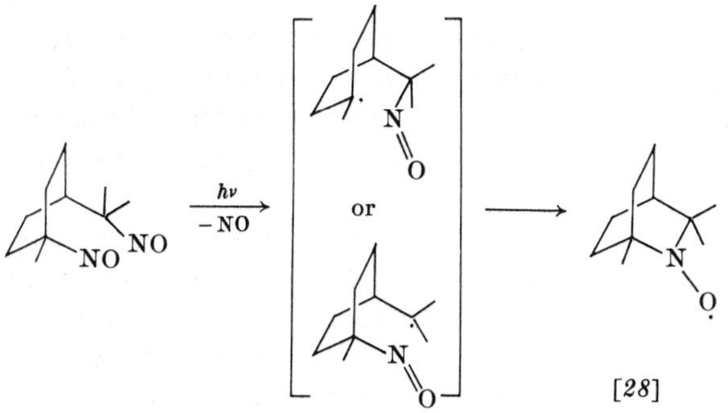

(5)

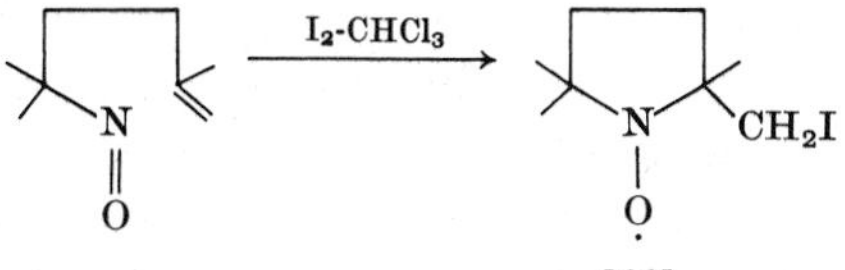

[*28*]

(6)

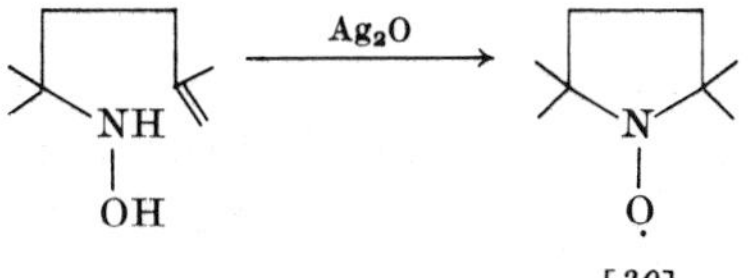

[*29*]

(7)

[*30*]

Spin-labeling experiments have usually involved derivatives [*31*] of 2,2,5,5-tetramethylpyrrolidine nitroxide or the piperidine analogs [*32*]. These stable radicals have been employed extensively as probes for the investigation of a range of diamagnetic systems by ESR, and they have

[50] A. Rassat and P. Rey, *Tetrahedron* **28**, 741 (1972).
[51] A. Rassat and P. Rey, *J. Chem. Soc. D* 1161 (1971).
[52] W. B. Motherwell and J. S. Roberts, *J. Chem. Soc. Chem. Commun.* 328 (1972).

been attached, commonly through standard reactions of appropriate functional groups (which do not affect the nitroxide moiety), to a variety of biological[39] and other macromolecules. Advantages inherent in the technique of spin labeling compared with other "tagging" methods are the sensitivity offered by ESR, the susceptibility of the

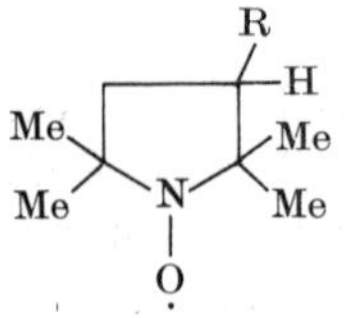

[31] 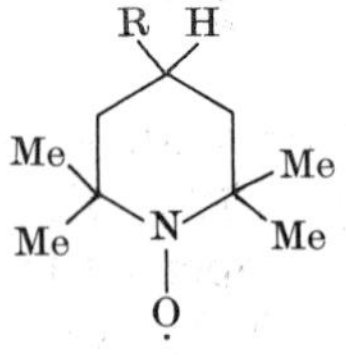   [32]

probe to the influence of its local environment (the nitrogen splitting is markedly affected by the polarity of its surroundings), the effect on the linewidth of the rotational mobility of the radical (broad lines result if the rate of tumbling is slow), and, usually, the absence of interfering signals from surrounding molecules.

To illustrate the scope of this technique, we choose a single example which involves the synthesis of a labeled coenzyme analog and an investigation of its binding to an enzyme. The probe is the adenosine diphosphate derivative (32, R = O-ADP) whose ESR signal in aqueous solution comprises a sharp, three-line spectrum characteristic of isotropic interaction between the unpaired electron and the ^{14}N nucleus.[53] In the presence of liver alcohol dehydrogenase this signal decreases in magnitude as the labeled substrate is bound to the enzyme; a broader signal also appears which is characteristic of the bound radical, now tumbling much more slowly since it is firmly attached to the macromolecule. Detailed examination of the results suggests[53] that two different types of binding site are available, and binding constants for both of them can be obtained. In some other cases,[39] detailed analysis of broadline spectra resulting from attachment of nitroxide moieties to macromolecules can give rotational correlation times for the radicals (this is related to the rate of tumbling and depends on the freedom of motion of the radical and the size of the macromolecule).

It has been suggested that some of the nitroxides discussed earlier might also find useful employment as spin labels. This category includes suitable derivatives of small-ring nitroxides, because of their compactness and inflexibility,[42] and also certain isoquinuclidine nitroxides which are rigid and more nearly spherical than the nitroxides

[53] H. Weiner, *Biochemistry* 8, 526 (1969).

thus far used. Two other types of stable radical reported are 33^{54} and 34 (X = O, H_2)^{55}; it has been suggested^{54} that radicals such as 33 may find application as phosphate spin labels.

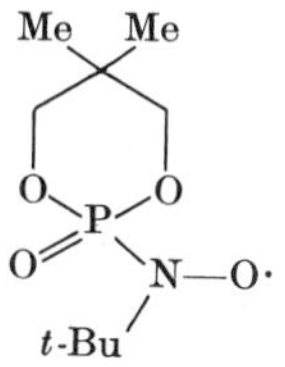

[33] 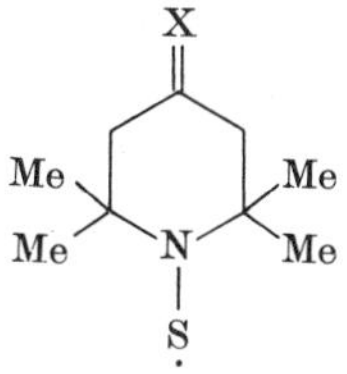   [34]

Another stable set of radicals are the nitronyl nitroxides (see Ullman et al.^{56} and preceding papers in this series). An example for consideration here is the radical analog [35] of histidine^{57}; in this case

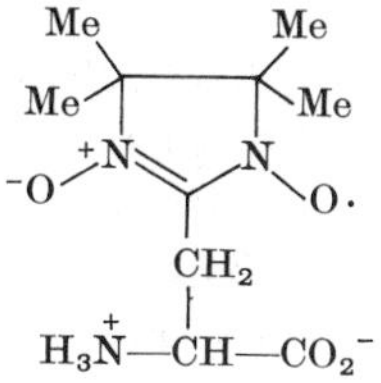

[35]

the methylene hydrogen splittings are nonequivalent due to the presence of an adjacent chiral center, and it has been pointed out that the difference between the splittings, Δa, should be sensitive to conformational preferences. This has been confirmed for a variety of radicals related to 35 (e.g., with substituents in the amino group), and in some instances line broadening suggests that conformational interconversion has become somewhat slowed; in certain cases the pH dependence of the spectra provides evidence for ion–dipole interaction between —$CO_2{}^-$ and the nitronyl nitroxide ring. The technique has been extended^{58} to some spin-labeled polypeptides in the angiotensin series for probing ion–dipole interactions and for the measurement of activation energies for conformational interconversion.

^{54} D. Gagnaire, A. Rassat, J. B. Robert, and P. Ruelle, *Tetrahedron Lett.* 4449 (1972).
^{55} J. E. Bennett, H. Sieper, and P. Tavs, *Tetrahedron* **23**, 1697 (1967).
^{56} E. F. Ullman, J. H. Osiecki, D. G. B. Boocock, and R. Darcy, *J. Amer. Chem. Soc.* **94**, 7049 (1972).
^{57} R. J. Weinkam and E. C. Jorgensen, *J. Amer. Chem. Soc.* **93**, 7028 (1971).
^{58} R. J. Weinkam and E. C. Jorgensen, *J. Amer. Chem. Soc.* **93**, 7033 (1971).

C. Heteroaliphatic Ion Radicals

It is perhaps not surprising that in only a few cases, compared with aromatic examples, has it proved possible to generate fairly stable anion radicals or cation radicals from heteroalicyclic compounds for ESR study. The fulfillment of certain special electronic requirements and careful experimental methods are required, as in the generation of the perfluoroazoalkane anion radicals 36[59] and 37[60] by electrochemical reduction (splittings, in mT, refer to solutions in N,N-dimethylformamide, DMF). Spin polarization and spiroconjugation mechanisms for

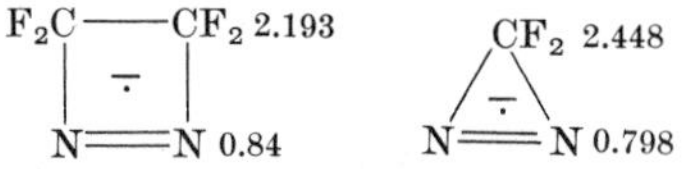

$$F_2C \overline{\hspace{1em}} CF_2 \ 2.193 \qquad CF_2 \ 2.448$$

(structures with $N{=\!=}N$ 0.84 for [36] and $N{=\!=}N$ 0.798 for [37])

[36]   [37]

unpaired electron delocalization on to fluorine atoms have been discussed for 37.[60]

Electrochemical oxidation of triethylenediamine in flowing solutions of acetonitrile leads to an ESR signal from the corresponding cation radical [38] which has splittings from two nitrogen atoms (1.696 mT each) and twelve hydrogens (0.734 mT).[61] The equivalence of the two nitrogen splittings indicates some transannular interaction (this may account for the unexpected stability of the radical), and the magnitude of $a(N)$ is in accord with the pyramidal nature of the radical center. The detection of stable cation radicals from 39 (X = NPh) produced[62]

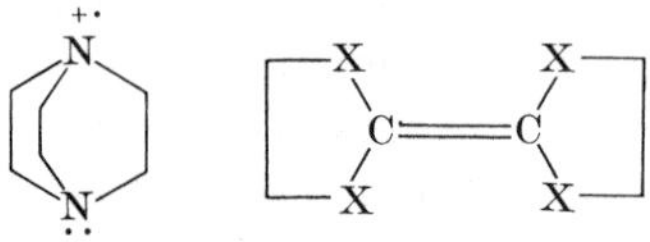

[38]   [39]

with O_2-MeCN, I_2, or Ag^+, and 39 (X = S), generated[63] electrolytically, is perhaps not surprising. Electrochemical oxidation of some cyclic[64] and bicyclic[65] dimethyldialkylhydrazines to stable cation radicals has

[59] J. L. Gerlock, E. G. Janzen, and J. K. Ruff, *J. Amer. Chem. Soc.* **92**, 2558 (1970).
[60] G. A. Russell, J. L. Gerlock, and G. R. Underwood, *J. Amer. Chem. Soc.* **94**, 5209 (1972).
[61] T. M. McKinney and D. H. Geske, *J. Amer. Chem. Soc.* **87**, 3013 (1965).
[62] D. M. Lemal and K. I. Kawano, *J. Amer. Chem. Soc.* **84**, 1761 (1962).
[63] J. Q. Chambers, N. D. Canfield, D. R. Williams, and D. L. Coffen, *Mol. Phys.* **19**, 581 (1970).
[64] S. F. Nelsen and P. J. Hintz, *J. Amer. Chem. Soc.* **93**, 7105 (1971).
[65] S. F. Nelsen and P. J. Hintz, *J. Amer. Chem. Soc.* **92**, 6215 (1970).

also been reported and the results have been interpreted in terms of conformational preferences and long-range splittings.

III. HETEROAROMATIC RADICALS

A. Anion Radicals

Many heteroaromatic compounds which have extended conjugation or which contain electron-withdrawing groups are susceptible to one-electron reduction, either chemically (e.g., with an alkali metal) or electrochemically, to give anion radicals which are often quite stable and amenable to ESR investigation. Interest has centered mainly on the interpretation of these spectra in terms of the unpaired electron distribution and on the comparison of these findings with the predictions of molecular orbital (MO) theory. The enormous number of relevant references forces us to be selective rather than comprehensive; the reader is referred elsewhere[9] for a detailed coverage of the necessary background to the molecular orbital approaches and of some earlier work on aromatic heterocyclic radicals (prior to 1967).

After many unsuccessful attempts to generate the pyridine anion radical [40] from the parent compound (the reduction of pyridine in ethereal solvents with an alkali metal leads to ESR signals from the 4,4′-bipyridyl anion radical; the early work has been reviewed[66]), this species was eventually prepared by continuous electroreduction in liquid ammonia at −75°C.[66] The hydrogen and nitrogen splittings (in mT) are as indicated. The ESR spectra of the anion radicals of pyrazine,

[40]

pyrimidine, some methylated pyridines, and several N oxides were also recorded. A number of Hückel and McLachlan[67] calculations were performed using suitable parameters for N and C—N ($\alpha_N = \alpha + 0.8\beta$; $\beta_{CN} = 1.076\beta$) and good agreement was obtained with "experimental" spin densities derived using hydrogen splittings in conjunction with the McConnell equation [Eq. (2)]. For a nitrogen atom in a π electron system like this, the splitting $a(^{14}N)$ should depend on the spin density at adjacent π centers as well as that on nitrogen (though ρ_N is thought to

[66] C. L. Talcott and R. J. Myers, *Mol. Phys.* **12**, 549 (1967).
[67] A. D. McLachlan, *Mol. Phys.* **3**, 233 (1960).

115

be usually dominant). For *40* the nitrogen splitting should then be given by the following expression:

$$a(N) = Q_N^N \rho_N - Q_{CN}^N(\rho_{C2} + \rho_{C6}) \tag{8}$$

where the spin densities, ρ, are evaluated using McConnell's equation. Inspection of the spectra for a variety of related anion radicals leads to best agreement with Eq. (8) for $Q_N^N = +2.73$ and $Q_{CN}^N = -0.17$ mT.[66]

Table I shows the calculated spin densities for the pyridine anion (obtained using the McLachlan modification of the Hückel method and the MO parameters discussed above) and those derived from the splittings using appropriate Q values. The results of a simple Hückel calculation are also included. Other authors have arrived at essentially

TABLE I

EXPERIMENTAL AND THEORETICAL SPIN DENSITIES FOR THE PYRIDINE ANION RADICAL [*40*]

Ring position	Observed splitting (mT)[a]	Experimental[a,b]	Spin density	
			Theoretical	
			Hückel model[c,d]	McLachlan model[a,e]
1	0.628(N)	0.247	0.258	0.275
2	0.355(H)	0.145	0.143	0.153
3	0.082(H)	0.033	0.059	0.007
4	0.970(H)	0.395	0.338	0.403

[a] C. L. Talcott and R. J. Myers, *Mol. Phys.* **12**, 549 (1967).
[b] Calculated from observed splittings using $Q^H = -2.45$, $Q_N^N = 2.73$, $Q_{CN}^N = -0.17$ mT in conjunction with Eqs. (2) and (8).
[c] A. R. Buick, T. J. Kemp, G. T. Neal, and T. J. Stone, *J. Chem. Soc. A* 1609 (1969).
[d] $\alpha_N = \alpha + 0.8\beta$, $\beta_{CN} = 1.1\beta$.
[e] $\lambda = 0.75$, $\alpha_N = \alpha + 0.8\beta$, $\beta_{CN} = 1.076\beta$.

similar conclusions,[9,68,69] in one case from a detailed study of a variety of anion radicals containing two pyridinyl nitrogen atoms.[68] Because Q_N^N is considerably greater than Q_{CN}^N, the simplified equation [Eq. (9)] (with $Q^N \sim 2.7$ mT) is sometimes used to derive spin densities on nitrogen from ^{14}N splittings:

$$a(N) = Q^N \rho_N \tag{9}$$

[68] J. C. M. Henning, *J. Chem. Phys.* **44**, 2139 (1966).
[69] F. Gerson, *Mol. Phys.* **24**, 445 (1972).

Equation (9) appears to prove realistic unless the nitrogen atom is flanked by two positions with considerable spin density.

Subsequently, methods used for the preparation of *40* have included sodium metal reduction[70] in hexamethylphosphoramide (HMPA) at 20°C and reduction of pyridine[71] (or reductive dehalogenation of some halogenated pyridines[72]) in a rapid-mixing flow system using solvated electrons from the dissolution of sodium in liquid ammonia.

A mechanism has been proposed[73] for the further reaction of *40* in anhydrous pyridine or in ethereal solvents to give the 4,4′-bipyridyl anion radical [Eqs. (10)–(12)]. In HMPA the extent of initial ion pairing, and therefore of dimerization, is minimized.

$$3\,Na + 3\,py \longrightarrow 3\,(Na^+, py^{\cdot-}) \tag{10}$$

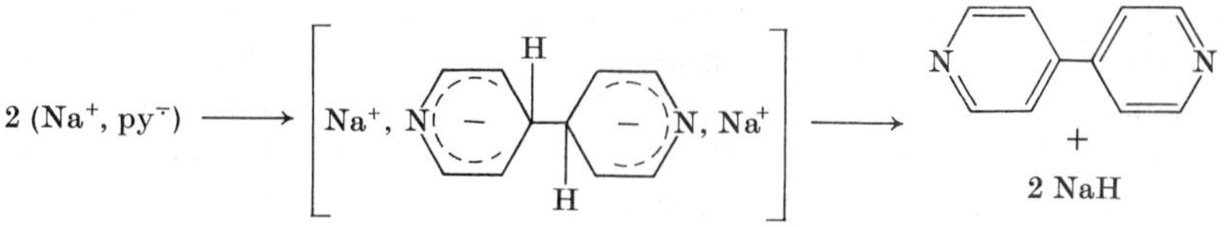

$$(11)$$

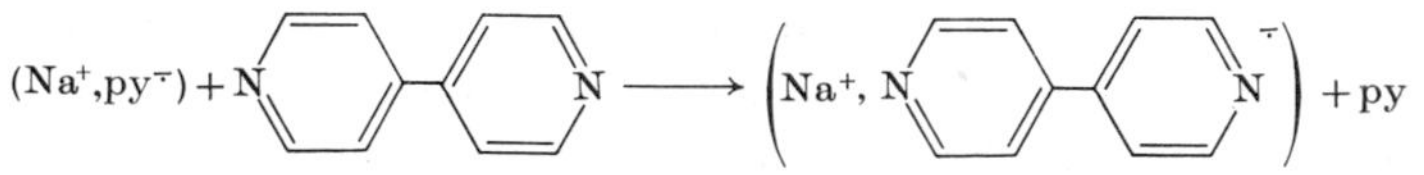

$$(12)$$

Extensive investigations have been made of anion radicals from pyridine derivatives including those of 3-nitropyridine (generated electrolytically),[74] 4-cyanopyridine (with potassium in 1,2-dimethoxyethane, DME),[75] methyl isonicotinate (electrochemically and with potassium),[76] both 2- and 4-vinylpyridine ($Na–NH_3$),[77] and some methylated pyridines (electrochemically and with alkali metals).[66,78] The reader is referred to these studies for MO calculations and for discussions of spin density distributions; in one case,[75] contact and solvent-separated ion pairs were distinguished. Intermediate Neglect of

[70] J. Chaudhuri, S. Kume, J. Jagur-Grodzinski, and M. Szwarc, *J. Amer. Chem. Soc.* **90**, 6421 (1968).

[71] A. R. Buick, T. J. Kemp, G. T. Neal, and T. J. Stone, *J. Chem. Soc. A* 1609 (1969).

[72] A. R. Buick, T. J. Kemp, G. T. Neal, and T. J. Stone, *J. Chem. Soc. A* 666 (1969).

[73] C. D. Schmulbach, C. C. Hinckley, and D. Wasmund, *J. Amer. Chem. Soc.* **90**, 6600 (1968).

[74] P. T. Cottrell and P. H. Rieger, *Mol. Phys.* **12**, 149 (1967).

[75] R. F. Adams, N. M. Atherton, A. E. Goggins, and C. M. Goold, *Chem. Phys. Lett.* **1**, 48 (1967).

[76] M. Hirayama, *Bull. Chem. Soc. Jap.* **40**, 1822 (1967).

[77] A. R. Buick, T. J. Kemp, and T. J. Stone, *J. Phys. Chem.* **74**, 3439 (1970).

[78] N. M. Atherton, F. Gerson, and J. N. Murrell, *Mol. Phys.* **5**, 509 (1962).

Differential Overlap calculations on a variety of pyridine anion radicals have also been reported.[79]

Other anion radicals with a single pyridinyl nitrogen include those of quinoline [41] and isoquinoline,[80] prepared by alkali metal reduction in DME at room temperature, and acridine [42] generated by electrochemical and alkali metal reduction.[70,81] Assistance in the assignment of all the indicated hydrogen and nitrogen splittings in 41 was provided

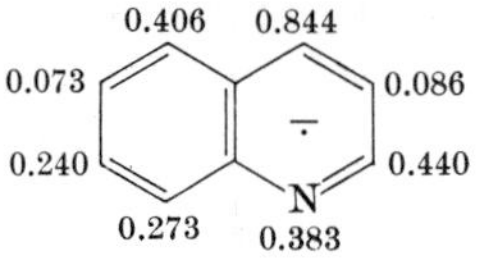

[41] 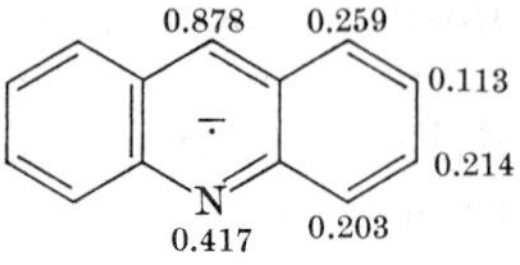   [42]

by comparison with data for the anions of all seven monomethyl derivatives.[80] In cases of low symmetry such as this it was pointed out that such a procedure is more reliable than a complete acceptance of the predictions of molecular orbital methods (even if suitable choices of h_N, k_{CN}, and Q can be made so that numerical agreement is achieved). Reduction of acridine with alkali metals in ethereal solvents can lead to dimerization[82,83] and, when the metal is in excess, to a dianion.[83]

Other anion radicals where a nitrogen atom is involved in a hybridized state different from those discussed above are the cycl[3.2.2]-azine anion [43],[84] some 1,6-bridged [10]annulene anion radicals (44, X = NH, NCH₃),[85] and the anion radical of N-isopropylcarbazole.[86] For 43 the splittings depend on whether sodium or potassium is used for the reduction; this may be a consequence of ion pairing. Molecular orbital calculations give good agreement with the observed splittings.[84] The hyperfine splittings for the ring hydrogens in 44 (X = NH) are

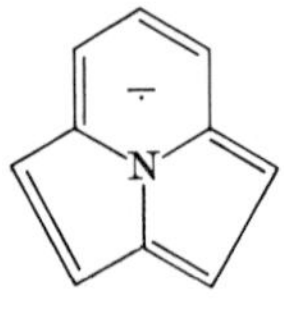

[43] 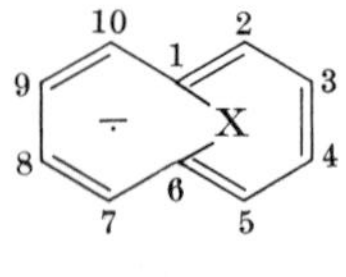   [44]

[79] D. M. Hirst, Theor. Chim. Acta 20, 292 (1971).
[80] L. Lunazzi, A. Mangini, G. F. Pedulli, and F. Taddei, J. Chem. Soc. B 163 (1970).
[81] S. Konishi, S. Niizuma, and M. Koizumi, Bull. Chem. Soc. Jap. 43, 3358 (1970); see also H. G. Hoeve and W. A. Yeranos, Mol. Phys. 12, 597 (1967).
[82] A. Carrington and J. dos Santos-Veiga, Mol. Phys. 5, 21 (1962).
[83] S. Niizuma, M. Okuda, and M. Koizumi, Bull. Chem. Soc. Jap. 41, 795 (1968).
[84] N. M. Atherton, F. Gerson, and J. N. Murrell, Mol. Phys. 6, 265 (1963); see also F. Gerson and J. D. W. van Voorst, Helv. Chim. Acta 46, 2257 (1963).

intermediate in magnitude between those for the oxygen- and methano-bridged analogs. As expected from MO considerations, the nitrogen splitting is very small (probably less than 0.005 mT); the hydrogen splittings are 0.328 and 0.286 mT (positions 2 and 10), 0.028 and 0.014 mT (3 and 9), and 0.058 mT (NH). When X = NMe, the radical decays to the azulene anion radical, whereas in the other cases the naphthalene anion results. The effect on hydrogen splittings of non-planarity in the π perimeter of bridged annulene anion radicals has been discussed.[87]

Anion radicals containing two pyridine-like nitrogen atoms include the thoroughly investigated pyrazine negative ion; this has been prepared[66] electrolytically in NH_3 (as has that of pyrimidine) and by alkali metal reduction[82,88] in DME and THF (in some cases metal splittings reveal the presence of ion pairs; anion radicals of 4,4'-bipyridyl, quinoxaline, phenazine, 1,4,5,8-tetraazaanthracene,[82] and pyridazine[88] have also been prepared in this way). Electrolysis in acetonitrile has also been used[68] to prepare the anions of pyrazine, pyridazine, quinoxaline, 1,5-diazanaphthalene, and phenazine (see also Stone and Maki[89]). The data for 45,[68] 46,[66] and 47[68] given here refer to electrolytically prepared "free" anions and are in millitesla. The reader is referred to all these papers for correlations of observed splittings with those calculated using standard MO procedures. More sophisticated calculational methods have also been reported.[90]

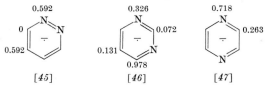

[45]   [46]   [47]

Considerable detailed information has resulted from studies of ion pair formation between the pyrazine negative ion and alkali metal cations in solvents of low dielectric constant[82,91-95]; the presence of

[85] F. Gerson, J. Heinzer, and E. Vogel, Helv. Chim. Acta 53, 95, 103 (1970).
[86] W. Klöpffer, G. Kaufmann, and G. Naundorf, Z. Naturforsch. A 26, 897 (1971).
[87] F. Gerson, K. Müllen, and E. Vogel, J. Amer. Chem. Soc. 94, 2924 (1972).
[88] R. L. Ward, J. Amer. Chem. Soc. 84, 332 (1962).
[89] E. W. Stone and A. H. Maki, J. Chem. Phys. 39, 1635 (1963).
[90] G. D. Zeiss and M. A. Whitehead, J. Chem. Soc. Faraday Trans. 2 526 (1972).
[91] J. dos Santos-Veiga and A. F. Neiva-Correia, Mol. Phys. 9, 395 (1965).
[92] N. M. Atherton and A. E. Goggins, Trans. Faraday Soc. 61, 1399 (1965).
[93] N. M. Atherton and A. E. Goggins, Trans. Faraday Soc. 62, 1702 (1966).
[94] N. M. Atherton, Trans. Faraday Soc. 62, 1707 (1966).
[95] C. A. McDowell and K. F. G. Paulus, Can. J. Chem. 43, 224 (1965).

associated species is revealed by splittings from metal nuclei, e.g., ^{39}K,[91] ^{23}Na,[92] and ^{7}Li.[93] Linewidth alternation effects for Na$^+$Pz$^-$ have been shown to be due to intramolecular cation exchange between positions of high charge density[92] (see also McDowell and Paulus[95]). It has been proposed,[93] on the basis of linewidth data and from the temperature dependence of the metal splittings, that for Li$^+$, Na$^+$, and K$^+$ the cation experiences a potential which has minima near the two nitrogen atoms, whereas for the larger Rb$^+$ and Cs$^+$ ions the effective potential has a single minimum over the center of the pyrazine ring. Factors which affect $a(M^+)$ have been discussed.[94,95] Reduction of pyrazine with sodium in the presence of sodium tetraphenylboride in THF leads to the formation of triple ions, Na$_2$Pz$^+$ [96]; from line broadening effects associated with the sodium splittings it was concluded that the sodium ions occupy positions close to the N—N axis of the ring system. For a series of alkali metal ions, the tightness of the binding within the triple ion and the magnitude of the spin density on the metal ion increase with decreasing size of the metal.[97]

The dimerization of pyrazine anion radicals, prepared at 21°C by reduction with potassium in DME, has been investigated using both ESR and electronic absorption spectroscopy[98]; kinetic parameters have been determined. Electroreduction of pyrimidine in DMF solution at −55°C leads[99] to the detection of the corresponding anion radical (with hyperfine splittings similar to those given above); continued electrolysis at −30°C generates another radical species whose structure was not assigned. However, 5-methylpyrimidine under similar conditions gives the 5,5′-dimethyl-2,2′-bipyrimidine anion radical, so it appears that dimerization may be the likely route. McLachlan MO calculations were reported for these radicals and also for some purine anion radicals; the ESR results should assist in elucidating the electronic properties of these biologically important systems.

Ion pairs formed between Group I metal ions and the 2,2′-bipyridyl anion radical have been detected[100–102] and the effects of coordination on the ESR parameters of the radical have been discussed in detail.[103]

[96] S. A. Al-Baldawi and T. E. Gough, *Can. J. Chem.* **48**, 2798 (1970).

[97] S. A. Al-Baldawi and T. E. Gough, *Can. J. Chem.* **49**, 2059 (1971).

[98] R. S. Hay and P. J. Pomery, *Aust. J. Chem.* **24**, 2287 (1971).

[99] M. D. Sevilla, *J. Phys. Chem.* **74**, 805 (1970).

[100] J. dos Santos-Veiga, W. L. Reynolds, and J. R. Bolton, *J. Chem. Phys.* **44**, 2214 (1966).

[101] E. König and H. Fischer, *Z. Naturforsch. A* **17**, 1063 (1962).

[102] A. Zahlan, F. W. Heineken, M. Bruin, and F. Bruin, *J. Chem. Phys.* **37**, 683 (1962).

[103] T. Takeshita and N. Hirota, *J. Amer. Chem. Soc.* **93**, 6421 (1971).

A spectrum attributed to the anion radical complex $Na[Cr(CO)_4$-2,2'-bipyridyl] has recently been reported.[104]

The ESR spectra of all six symmetrical diazanaphthalene anion radicals (from quinoxaline, phthalazine, and 1,5-, 1,8-, 2,6-, and 2,7-naphthyridine), generated by electrolytic reduction in DMF, have been recorded.[105] Theoretical spin densities were calculated by both the Hückel and the Pariser-Parr-Pople SCF methods.[106] In a similar study,[107] anion radicals of quinoxaline, 2,3-dimethylquinoxaline, 1,5-, 1,6-, 1,7-, 1,8-, 2,6-, and 2,7-naphthyridine, 3,7-dimethyl-1,5-naphthyridine, and 1-methyl-2,7-naphthyridine were prepared by potassium reduction of the parent heterocycles in DME and HMPA. Theoretical spin densities were derived using a number of standard MO procedures. For both investigations,[105,107] the simple Hückel approach seems to lead to good agreement.

Other anion radicals reported include those of 1,4,5,8-tetraazanaphthalene[108,109] and the 2,3,6,7- and 1,4,6,7-analogs[109] (by electrochemical and chemical methods; MO calculations were reported), that from 1,3,6,8-tetraazapyrene[110] (the corresponding cation radical was also produced), and those produced electrolytically[111] from 9,10-diazaphenanthrene, 2,2'-bipyrimidine, and $\Delta^{2,2'}$-biisobenzimidazolylidene (but see Kuhn et al.[112] and Russell et al.[113]).

Another method employed to generate anion radicals of some azo compounds and their vinylogs involves the disproportionation reaction between a suitable azo compound (R—N=N—R) and the related hydrazo species (R—NH—NH—R) in basic solution.[113] Anion radicals can also be generated in some cases by one-electron reduction of the azo compound with the propiophenone enolate anion. Spectra have been described for the anion radicals of phenazine, 2,3-diphenylquinoxaline, benzoxadiazole, benzo[c]cinnoline, and some tetrazines. Some 3,6-

[104] Y. Kaizu and H. Kobayashi, Bull. Chem. Soc. Jap. 45, 470 (1972).

[105] D. M. W. van den Ham, J. J. du Sart, and D. van der Meer, Mol. Phys. 21, 989 (1971).

[106] R. G. Parr, "Quantum Theory of Molecular Electronic Structure." Benjamin, New York, 1963.

[107] P. Cavalieri d'Oro, R. Danieli, G. Maccagnani, G. F. Pedulli, and P. Palmieri, Mol. Phys. 20, 365 (1971).

[108] F. Gerson and W. L. F. Armarego, Helv. Chim. Acta 48, 112 (1965).

[109] R. Danieli, L. Lunazzi, and G. Placucci, J. Amer. Chem. Soc. 93, 5850 (1971).

[110] F. Gerson, Helv. Chim. Acta 47, 1484 (1964).

[111] D. H. Geske and G. R. Padmanabhan, J. Amer. Chem. Soc. 87, 1651 (1965).

[112] R. Kuhn, P. Skrabal, and P. H. H. Fischer, Tetrahedron 24, 1843 (1968).

[113] G. A. Russell, R. Konaka, E. T. Strom, W. C. Danen, K.-Y. Chang, and G. Kaupp, J. Amer. Chem. Soc. 90, 4646 (1968).

dialkyltetrazines spontaneously undergo disproportionation in basic solution to give the corresponding anion radicals.[113,114]

The reader is referred to the original literature for details of the ESR spectra of the anion radicals of p-dicyanotetrazine,[115] 2,3,5,6-tetracyanopyridine and pentacyanopyridine,[116] and the N-oxides of a variety of pyridinyl derivatives.[117-119] Suitable Q values and MO parameters have been proposed. A wealth of detail exists on the stable anion radicals of phthalimide, phthalic anhydride, and some related radicals.[120]

Of the few examples of heterocyclic anion radicals containing nitrogen in combination with another heteroatom, the most studied have been the benzodiazole derivatives (48, X = oxygen,[121,122] sulfur,[121-123] and selenium[121-124]) prepared by electron transfer from carbanions[121] and with alkali metals.[122-124] Inspection of the splittings confirms that sulfur and selenium delocalize the unpaired electron to a greater extent

[48]

than does oxygen[121] and that MO treatments based on a p orbital model for S[121,122] and Se[124] appear satisfactory. The perfluoro-2,1,3-benzoselenadiazole anion radical[125] and some naphthothiadiazole anion radicals[122] have also been reported.

Among the anion radicals containing Group VI elements are those from dibenzofuran[126-128] and its sulfur[126,127,129] and selenium ana-

[114] H. Malkus, M. A. Battiste, and R. M. White, *J. Chem. Soc. D* 479 (1970).

[115] A. Carrington, P. Todd, and J. dos Santos-Veiga, *Mol. Phys.* **6**, 101 (1963).

[116] M. T. Jones, *J. Amer. Chem. Soc.* **88**, 5060 (1966).

[117] E. G. Janzen and J. W. Happ, *J. Phys. Chem.* **73**, 2335 (1969).

[118] L. Lunazzi, A. Mangini, G. Placucci, and F. Taddei, *J. Chem. Soc. B* 440 (1970).

[119] T. Kubota, K. Nishikida, H. Miyazaki, K. Iwatani, and Y. Oishi, *J. Amer. Chem. Soc.* **90**, 5080 (1968).

[120] S. F. Nelsen, *J. Amer. Chem. Soc.* **89**, 5256, 5925 (1967); D. C. McCain, *J. Magn. Resonance* **7**, 170 (1972); M. Hirayama, *Bull. Chem. Soc. Jap.* **40**, 1557 (1967); R. E. Sioda and W. S. Koski, *J. Amer. Chem. Soc.* **89**, 475 (1967).

[121] E. T. Strom and G. A. Russell, *J. Amer. Chem. Soc.* **87**, 3326 (1965).

[122] N. M. Atherton, J. N. Ockwell, and R. Dietz, *J. Chem. Soc. A* 771 (1967).

[123] M. Kamiya and Y. Akahori, *Bull. Chem. Soc. Jap.* **43**, 268 (1970).

[124] M. F. Chiu and B. C. Gilbert, *J. Chem. Soc. Perkin Trans. 2* 258 (1973).

[125] J. Fajer, B. H. J. Bielski, and R. H. Felton, *J. Phys. Chem.* **72**, 1281 (1968).

[126] R. Gerdil and E. A. C. Lucken, *J. Amer. Chem. Soc.* **87**, 213 (1965).

[127] F. C. Adam and C. R. Kepford, *Can. J. Chem.* **49**, 3529 (1971).

[128] A. G. Evans, P. B. Roberts, and B. J. Tabner, *J. Chem. Soc. B* 269 (1966).

logs.[124,126] The hydrogen splittings (mT) for *49* and *50* refer to solutions of the anion radicals generated in DME at $-40°C$ by potassium metal reduction; the assignments have been justified.[127] Molecular orbital calculations suggest that a p orbital model is appropriate both for S,[126]

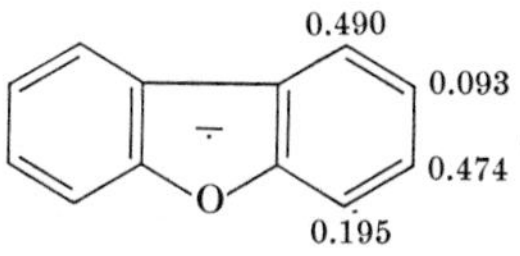

[*49*] 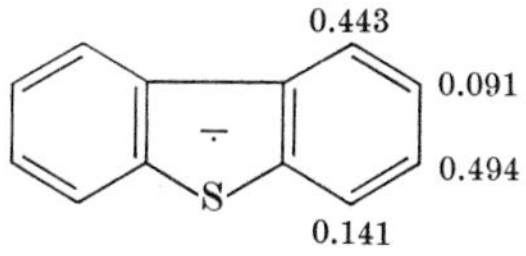   [*50*]

and for Se[124] and, in another study, it has been pointed out[127] that comparison of the spectra from these "bridged" biphenyl anion radicals with that from biphenyl itself is instructive. Further reactions of the dibenzofuran anion radical have been followed by optical and ESR spectroscopy[128]; kinetic parameters for ring opening via C—O cleavage have been determined for a variety of gegen-ions, and the rate increases by a factor of 25 from Cs^+ to Li^+ at 40°C. The ESR spectra of the anion radicals of dibenzo[*b,f*]thiepin and some methylated derivatives have also been recorded.[130]

Although the anion radical of dibenzothiophene is fairly stable, attempts to prepare that of thiophene itself have thus far proved unsuccessful. Recently, however, condensed species such as *51* and some related compounds, e.g., *52*, have attracted attention.[131] These radicals, which exhibit interesting linewidth effects stemming from anisotropy, were prepared at $-100°C$ by reduction of the parent compounds with sodium–potassium alloy, and the splittings were assigned on the basis

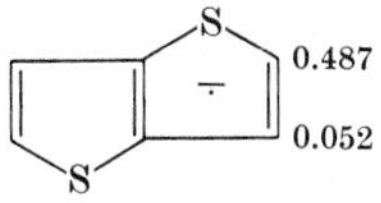

[*51*] 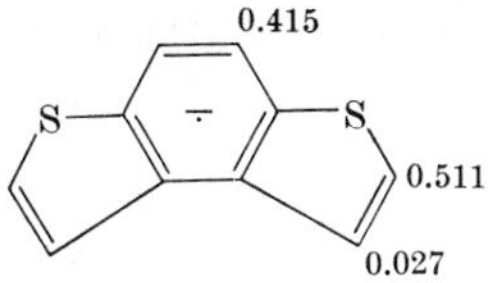   [*52*]

of MO calculations (p orbital participation for the S atoms, rather than d orbital participation, provided the best model). It has been pointed out[131] that the stability of the thiophene anion radical is less than that of benzene. Further, the extra stability of the anion radicals of the corresponding noncyclized analogs, the dithienylethylenes (e.g., *53*),[132]

[129] D. H. Eargle, Jr., and E. T. Kaiser, *Proc. Chem. Soc.* 22 (1964).

[130] M. M. Urberg and E. T. Kaiser, *J. Amer. Chem. Soc.* **89**, 5931 (1967).

[131] L. Lunazzi, G. Placucci, M. Tiecco, and G. Martelli, *J. Chem. Soc. B* 1820 (1971).

[132] L. Lunazzi, A. Mangini, G. Placucci, P. Spagnolo, and M. Tiecco, *J. Chem. Soc. Perkin Trans. 2* 192 (1972).

indicates that annelation has a marked destabilizing effect.[131] The radical *53* has been generated[132] from *trans*-1,2-di(2-thienyl)ethylene by reduction with potassium at − 80°C. At this temperature the presence of two species is revealed by ESR; the major isomer (80%) was tentatively assigned the S-*cis*,*cis* structure [*53*] with the coupling constants indicated, and the less stable isomer the S-*cis*, *trans* conformation. The g values of the two conformers are noticeably different ($\Delta g = 5 \times 10^{-3}$). Calculations within the p model framework were employed

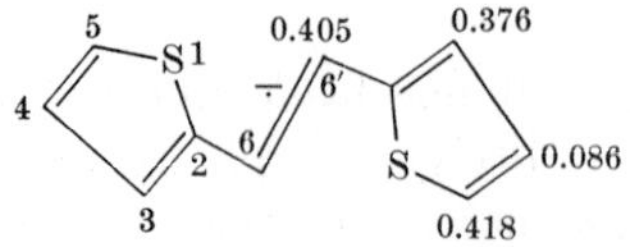

[*53*] 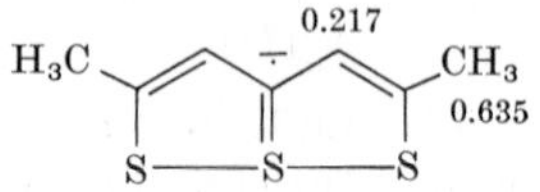   [*54*]

successfully, the hindered rotation of the 2-thienyl groups being accommodated by the variation of the resonance integrals for the 2—6 and 2′—6′ bonds ($\beta = \beta_0 \cos \theta$). The best fit with the experimental data was obtained for a planar arrangement. Different rotamers were also observed for the 2,2′-dithienyl anion radical, generated with potassium in DME, at − 80°C.[133] Other interesting condensed systems which have yielded fairly stable anion radicals include 2,5-dimethyl-6*a*-thiathiophthene[134] (accordance has been obtained between the proposed MO model and the splittings for *54*, generated electrolytically in DMF) and [18]annulene-1,4;7,10;13,16-trisulfide.[135] The radical [*55*] from the latter prepared by alkali metal reduction in DME or THF[135] has a lifetime of several hours at room temperature and a spectrum which shows six pairs of hydrogen splittings (and $g = 2.00378$). The coupling constants were assigned on the basis of calculated spin populations; the

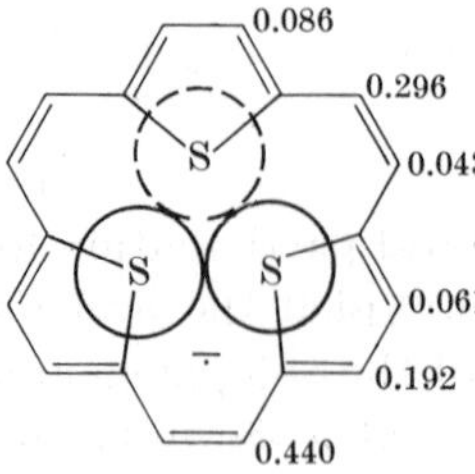

[*55*]

[133] P. Cavalieri d'Oro, A. Mangini, G. F. Pedulli, P. Spagnolo, and M. Tiecco, *Tetrahedron Lett.* 4179 (1969).

[134] F. Gerson, R. Gleiter, J. Heinzer, and H. Behringer, *Angew. Chem. Int. Ed.* **9**, 306 (1970).

[135] F. Gerson and J. Heinzer, *Helv. Chim. Acta* **51**, 366 (1968).

lack of threefold symmetry observed seems to be due to the long life ($> 10^{-7}$ sec) of the conformation indicated, in which only two of the three sulfur atoms lie on the same side of the radical. The relative ease of generating anion radicals from ring-substituted thiophenes is due evidently to the stabilizing effect of suitable substituents, e.g., nitro and carbonyl. For example, photolysis of thiophene-2-aldehyde, thiophene-3-aldehyde, 2-acetylthiophene, and 2,2'-dithienone in solutions of sodium methoxide in methanol has been used to prepare the respective anion radicals.[136] With 2-acetylthiophene, the ESR spectra reveal two species whose relative concentrations are temperature dependent and which appear to be the rotational isomers 56 and 57; calculations suggest that a p model for sulfur gives slightly better agreement than a d model, and measurements of radical concentrations as a function of temperature yield $\Delta H = 2.9 \pm 0.4$ kJ mole^{-1} for the interconversion. More recently,[137] a conformational analysis of anion radicals from thiophene-2,5-dicarbaldehyde and some related

[56]   [57]

derivatives has been undertaken. Other radicals from thiophene derivatives whose ESR spectra have been described include some thienyl phenyl ketone anion radicals, generated by electrolysis,[138] and some anion radicals generated similarly from nitrothiophenes[139] (where both p and d models for sulfur lead to reasonable agreement with experiment). The semidiones RC(O·)=C(O$^-$)R, with R = 2-thienyl, 3-thienyl, and 2-furyl, have been described[140]; the resonance stabilizations of the 2-thienyl- and 2-furylmethyl radicals (estimated from the reactivity of some methylthiophenes, methylfurans, methylbenzothiophenes, and methylbenzofurans toward phenyl radicals) have been correlated with the delocalization indicated by the ESR spectra of the semidiones (see also p. 137). Electron spin resonance spectra have been reported for the 2- and 3-thienylmethyl radicals, generated from methyl-substituted

[136] A. Hudson and J. W. E. Lewis, *Tetrahedron* **26**, 4413 (1970).

[137] L. Lunazzi, G. F. Pedulli, M. Tiecco, C. Vincenzi, and C. A. Veracini, *J. Chem. Soc. Perkin Trans. 2* 751 (1972).

[138] L. Kaper, J. U. Veenland, and Th. J. de Boer, *Spectrochim. Acta Part A* **24**, 1971 (1968).

[139] E. A. C. Lucken, *J. Chem. Soc. A* 991 (1966).

[140] E. T. Strom, G. A. Russell, and J. H. Schoeb, *J. Amer. Chem. Soc.* **88**, 2004 (1966).

thiophenes with ·O-t-Bu (from photolysis of di-t-butyl peroxide)[136,141]; in each case, nonequivalent methylene hydrogen splittings indicate restricted rotation.

The majority of ESR investigations of oxygen-containing anion radicals has been concerned with fused dibenzo systems. There has also been wide interest in the corresponding sulfur radicals, and comparative studies are common. Xanthene undergoes electrolytic reduction in DMF at 10°C to give an ESR spectrum characterized by only four triplet splittings and attributed to the xanthene-9-one ketyl [58][142]; the splittings are assigned on the basis of MO calculations. Thioxanthene behaves similarly. The spectra of 58 and its analog from thioxanthene-

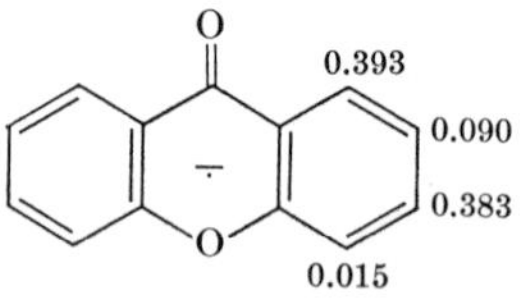

[58] 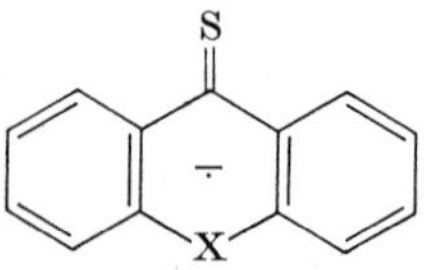   [59]

9-one have also been obtained together with those from the anion radicals 59 (X = O, S) by sodium metal reduction of the appropriate parent molecules in various ethereal solvents.[143] Free ions and ion pairs have been generated and the higher g values for the free ions than the corresponding ion pairs indicate that more spin density exists in the carbonyl or thiocarbonyl group in the nonassociated species. Molecular orbital calculations give good agreement.

Considerable attention has been paid to the investigations of anion radicals of aromatic compounds containing sulfone and sulfoxide groupings. That the latter groups are more effective than a sulfur atom for delocalizing the unpaired electron is revealed by the observation[129] that the spectrum width (effectively a measure of the unpaired spin density in the rings) is much smaller for 60 (X = SO) than for 60 (X = S); for 60 (X = SO$_2$) only a small further decrease is observed. Similar effects were noted for a variety of radicals [61] with X = CO,

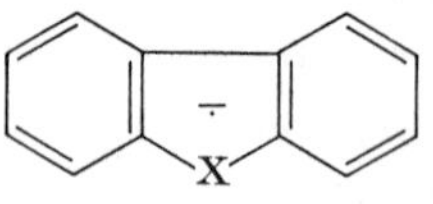

[60] 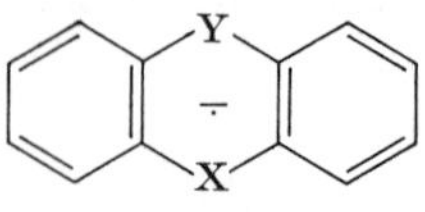   [61]

[141] A. Hudson, H. A. Hussain, and J. W. E. Lewis, *Mol. Phys.* **16**, 519 (1969).
[142] B. J. Tabner and J. R. Zdysiewicz, *J. Chem. Soc. B* 1659 (1971).
[143] L. J. Aarons and F. C. Adam, *Can. J. Chem.* **50**, 1390 (1972).

$Y = S, SO_2$[144]; $X = S, Y = SO$; $X = SO, Y = SO_2$; and $X = Y = SO, SO_2$.[145] Anion radicals of *cis*- and *trans*-thianthrene-5,10-dioxide have been characterized.[145] Electron spin resonance spectra from other sulfone anion radicals have been recorded[146,147] and MO parameters for the sulfone group have been proposed (e.g.,[147] $\alpha_{SO_2} = \alpha - 2.0\beta$; $\beta_{C-SO_2} = 0.8\beta$). In a recent study[148] of the thioxanthone sulfoxide anion radical and that from dibenzo[b,f]thiepin, the delocalizing ability of the sulfoxide group has been discussed and it has also been shown that conjugation through a sulfoxide bridge is less effective than through a vinyl residue. For *62*, prepared by potassium metal reduction[149] in DME at 25°C, two hydrogen splittings (0.0055 and 0.2110 mT, assigned

[62]

to H-1 and H-2, respectively) and three ^{13}C splittings (0.474, 0.1984, and 0.097 mT, assigned to C-2, C-1, and C-4a, respectively, on the basis of MO calculations) were observed. A p orbital model seems appropriate, and the inability to detect a ^{33}S splitting, together with the low g value (2.0023), suggests that little spin is associated with sulfur.

The reader is referred elsewhere for discussions of the chemistry and the ESR spectra of anion radicals from indigo, thioindigo, and related compounds,[150] from ninhydrin and alloxan,[151] and from substituted maleic anhydride derivatives, "semifuraquinones" (see Nelsen *et al.*[152] and earlier papers in the series).

[144] E. T. Kaiser and D. H. Eargle, Jr., *J. Amer. Chem. Soc.* **85**, 1821 (1963).
[145] E. T. Kaiser and D. H. Eargle, Jr., *J. Phys. Chem.* **69**, 2108 (1965); *J. Chem. Phys.* **39**, 1353 (1963).
[146] G. Vincow, *J. Chem. Phys.* **37**, 2484 (1962); J. P. Keller and R. G. Hayes, *ibid.* **46**, 817 (1967).
[147] R. Gerdil and E. A. C. Lucken, *Mol. Phys.* **9**, 529 (1965).
[148] A. Trifunac and E. T. Kaiser, *J. Phys. Chem.* **74**, 2236 (1970).
[149] D. H. Eargle, Jr., *J. Phys. Chem.* **73**, 1854 (1969).
[150] G. A. Russell, C. L. Myers, P. Bruni, F. A. Neugebauer, and R. Blankespoor, *J. Amer. Chem. Soc.* **92**, 2762 (1970); M. Bruin, F. Bruin, and F. W. Heineken, *J. Chem. Phys.* **37**, 135 (1962); G. A. Russell and R. Konaka, *J. Org. Chem.* **32**, 234 (1967).
[151] G. A. Russell and M. C. Young, *J. Amer. Chem. Soc.* **88**, 2007 (1966).
[152] S. F. Nelsen, E. F. Travecedo, and E. D. Seppanen, *J. Amer. Chem. Soc.* **93**, 2913 (1971).

The reaction of phenylbiphenylenephosphine with alkali metals evidently proceeds via phenyl cleavage, and ESR spectra of the anion radicals *63*, with $a(^{31}P) = 0.85$ mT, and *64*, with $a(^{31}P) = 0.88$ mT, were detected.[153] Reduction of some substituted phospholes with alkali

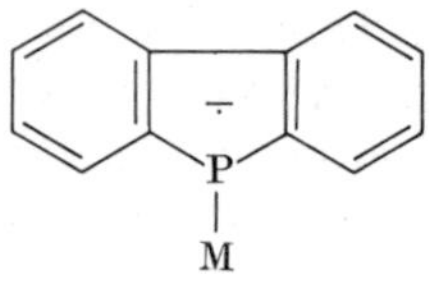

[*63*] 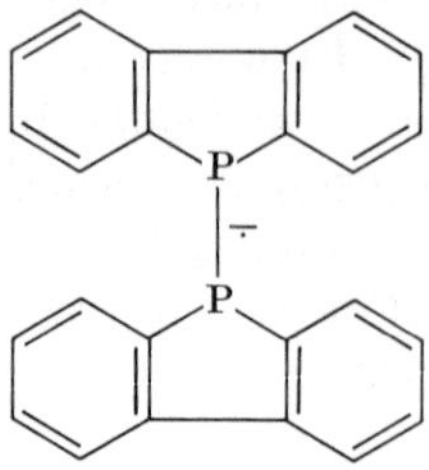   [*64*]

metals at low temperatures gives spectra attributed to the mono-anions.[154] At room temperature, reaction of some phenyl-substituted phospholes with potassium leads to a complex series of radicals resulting from cleavage of a phenyl group,[155] including radicals of the type R_2PK^-.[154] Reduction of some phosphole oxides with alkali metals has been used to generate the corresponding anion radicals, e.g., *65*.[156]

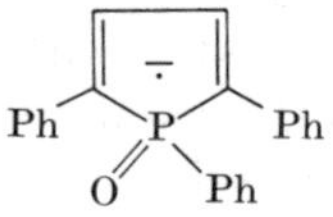

[*65*] 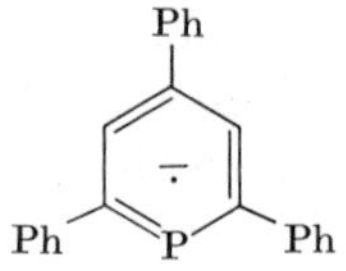   [*66*]

The splittings suggest that, as in the phosphole anions, there is delocalization around an aromatic phosphorus-containing ring; the dependence of $a(^{31}P)$ on the metal employed suggests that close association of the gegen ion with the phosphorus atom (possibly via d orbital interaction) takes place. Phosphorin anion (and also cation) radicals (e.g., *66*) have been described,[157,158] and $\sigma-\pi$ polarization parameters for ^{31}P have been determined.[158]

The preparation of the phosphorus-containing (neutral) spiro radical

[153] A. D. Britt and E. T. Kaiser, *J. Org. Chem.* **31**, 112 (1966).
[154] D. Kilcast and C. Thomson, *Tetrahedron* **27**, 5705 (1971).
[155] C. Thomson and D. Kilcast, *Angew. Chem. Int. Ed.* **9**, 310 (1970).
[156] C. Thomson and D. Kilcast, *J. Chem. Soc. D* 782 (1971).
[157] K. Dimroth and F. W. Steuber, *Angew. Chem. Int. Ed.* **6**, 445 (1967).
[158] C. Thomson and D. Kilcast, *J. Chem. Soc. D* 214 (1971).

[67][159] and the related anion [68],[160] has been reported. Other silicon-containing anion radicals include those of some substituted 1,4-disila-cyclohexadienes,[161] 5,10-dihydrosilanthrenes,[162] 1,1-dimethyl-1-sila-

[67]   [68]

cyclopentadienes,[163] and 9,9-dialkyl-9-silafluorenes.[164] For the latter, $d-\pi$ interactions have been proposed. Some heterocyclic radicals containing boron have also been reported.[165]

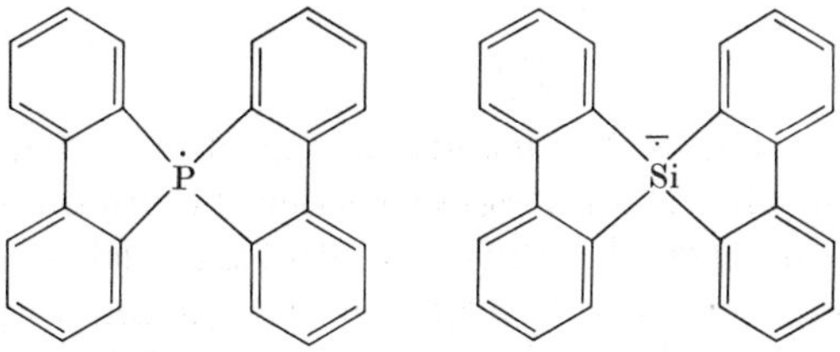

B. Cation Radicals

Stable cation radicals can often be prepared from compounds containing extended conjugation and electron-donating substituents; the methods used for one-electron oxidation include electrolysis, reaction with sulfuric acid, treatment with a suitable Lewis acid (e.g., $AlCl_3$ in nitromethane), and oxidation with a high-valent metal ion. As with the corresponding investigations of anion radicals, concern has normally been with detailed analysis of the spectra, assignment of the splittings to different nuclei in the radical, and comparison of "observed" unpaired electron distributions with those predicted using MO calculations.

The scope of this approach is nicely illustrated by the early report of the ESR spectra from cation radicals 69–71, generated in concentrated sulfuric acid solution.[166] Improved resolution proved possible when the parent heterocycles were oxidized with $AlCl_3$–CH_3NO_2,[167] and this allowed complete resolution of hydrogen and ^{33}S splittings (these are

[159] R. Rothuis, T. K. J. Luderer, and H. M. Buck, *Rec. Trav. Chim. Pays-Bas* **91**, 836 (1972).
[160] R. D. Cowell, G. Urry, and S. I. Weissman, *J. Amer. Chem. Soc.* **85**, 822 (1963).
[161] E. G. Janzen, J. B. Pickett, and W. H. Atwell, *J. Amer. Chem. Soc.* **90**, 2719 (1968).
[162] E. G. Janzen and J. B. Pickett, *J. Amer. Chem. Soc.* **89**, 3649 (1967).
[163] E. G. Janzen, J. B. Pickett, and W. H. Atwell, *J. Organometal. Chem.* **10**, P6 (1967).
[164] N. V. Eliseeva, T. L. Krasnova, E. A. Chernyshev, A. N. Pravednikov, and V. L. Rogachevskii, *Zh. Strukt. Khim.* **13**, 519 (1972); *Chem. Abstr.* **77**, 94963f (1972).
[165] G. R. Eaton, *Inorg. Nucl. Chem. Lett.* **8**, 634 (1972); M. A. Kuck and G. Urry, *J. Amer. Chem. Soc.* **88**, 426 (1966).
[166] E. A. C. Lucken, *Theor. Chim. Acta* **1**, 397 (1963).
[167] P. D. Sullivan, *J. Amer. Chem. Soc.* **90**, 3618 (1968).

0.984
S
1 2   0.282
+·
S
[69]

0.935
S
1 2   0.332
+·
6
0.1056   5
S
0.0201
[70]

0.915   0.0135
S
9   1 2   0.128
+·
S
[71]

shown, in mT). The MO calculations reported include those where, first, only p orbitals on sulfur were considered (with $\alpha_S = \alpha + 1.0\beta$ and $\beta_{CS} = 0.566\beta$) and then also where a d orbital model was assumed, with $\alpha_S = \alpha$, $\beta_{SS'} = \beta$, and $\beta_{CS} = 0.8\beta$; this was found to give somewhat better agreement.[166] For the second investigation, however, only a p orbital model (Hückel-McLachlan calculation) was employed ($h_S = 1.20$ and $k_{CS} = 0.65$ or 0.55); this gave excellent agreement for $a(H)$, and also for $a(^{33}S)$ using a simple proportionality constant between $\rho_S{}^\pi$ and the splitting ($Q^S = 3.34$ mT).[167] The results are summarized in Table II. It may be mentioned at this point that for most of the sulfur-containing cation radicals reported in this section, the p orbital model has usually been found to be adequate.

The extra stability afforded by the flanking aromatic rings in *71* accounts for the many examples of cation radicals of this type whose

TABLE II

EXPERIMENTAL AND CALCULATED SPLITTINGS (mT) FOR SOME SULFUR-CONTAINING CATION RADICALS

Radical	Position	Observed splitting[a]	Calculated splitting	
			p Model[a,b]	d Model[c]
1,4-Dithiin$^{+\cdot}$ [69]	1	0.984 (S)	0.965	— —
	2	0.282 (H)	(−)0.292	0.300[d] 0.300[e]
1,4-Benzodithiin$^{+\cdot}$ [70]	1	0.935 (S)	0.942	— —
	2	0.332 (H)	(−)0.436	0.316[d] 0.359[e]
	5	0.0201 (H)	(−)0.0210	0.047[d] 0.020[e]
	6	0.1056 (H)	(−)0.0635	0.098[d] 0.074[e]
Thianthrene$^{+\cdot}$ [71]	1	0.0135 (H)	(−)0.0103	0.037[d] 0.008[e]
	2	0.128 (H)	(−)0.117	0.106[d] 0.098[e]
	9	0.915 (S)	0.919	— —

[a] P. D. Sullivan, *J. Amer. Chem. Soc.* **90**, 3618 (1968).

[b] McLachan's method ($\lambda = 1.2$) with $h_S = 1.20$, $k_{CS} = 0.65$, $Q^S = 3.34$, $Q^H = (−)2.8$ mT.

[c] E. A. C. Lucken, *Theor. Chim. Acta* **1**, 397 (1963); negative signs for hydrogen splittings understood.

[d] Hückel's method with $\alpha_S = \alpha$, $\beta_{SS'} = \beta$, $\beta_{CS} = 0.8\beta$. $|Q^H| = 2.4$ mT.

[e] As d, but using McLachlan's method ($\lambda = 1.0$).

ESR spectra have been investigated; in this section we present a summary, rather than an exhaustive coverage, of the relevant findings. For example, the cation radicals of dibenzo-p-dioxin [72] and phenoxathiin [73] together with 71 have all been prepared from the parent compounds in 95% sulfuric acid.[168] The g values increase, as expected, along the series 72, 73, and 71; they are 2.0035, 2.0061, and 2.0081, respectively. g Values of 2.0228 and 2.0315 have been reported for the phenoxase-

[72]   [73]

lenin[124] and the dibenzo-p-diselenin cation radicals.[168] Molecular orbital calculations for the sulfur-containing radicals have been carried out using a p orbital model[168]; observed ^{13}C splittings for 72 were assigned to ring positions on the basis of calculated spin densities. Ionization potentials and electronic absorption spectra were also measured in order to obtain supplementary information about π-electron energy levels for these systems,[168] and the observed g values have been correlated with spin density distributions.[124,168] For the phenoxaselenin and dibenzo-p-diselenin cations the following MO parameters were employed[124] (p orbital model): $h_O = 2.0$, $k_{CO} = 0.8$; $k_{Se} = 1.21$, $k_{CSe} = 0.57$. For a variety of selenium-containing radicals, for which ^{77}Se splittings were detected, a simple proportionality between ρ_{Se}^{π} and $a(^{77}Se)$ seems to be obeyed, with $Q^{Se} = 12.3$ mT.[124]

Radical 72 and some derivatives have also been generated in trifluoromethanesulfonic acid solution,[169] a method which enables very narrow lines ($\Delta H < 0.004$ mT) to be obtained. The radical 73 has also been generated by electrochemical oxidation of the parent molecule in acetonitrile,[170] and some methylated and fluorinated derivatives have been prepared using a variety of oxidants including Fe(III) and Ti(IV)[171]; further MO calculations were reported. It has been found possible to isolate the thianthrene cation radical [71] as its trichlorodiiodide and perchlorate.[172] These salts are quite stable and have permitted the reactions of 71 with a variety of substrates (e.g., water,

[168] B. Lamotte and G. Berthier, *J. Chim. Phys. Physicochim. Biol.* **63**, 369 (1966).
[169] G. C. Yang and A. E. Pohland, *J. Phys. Chem.* **76**, 1504 (1972).
[170] C. Barry, G. Cauquis, and M. Maurey, *Bull. Soc. Chim. Fr.* 2510 (1966).
[171] M. Hillebrand, O. Maior, V. Em. Sahini, and E. Volanschi, *J. Chem. Soc. B* 755 (1969).
[172] Y. Murata and H. J. Shine, *J. Org. Chem.* **34**, 3368 (1969).

halide ions, and amines) to be investigated. For example, the perchlorate salt reacts with water to form equal amounts of thianthrene and thianthrene-5-oxide. These products arise not by direct reaction of *71* itself with water, but rather there is an initial disproportionation of the cation radical in which thianthrene and its dication are formed; the latter then reacts with water to yield the 5-oxide.

The cation radicals *74* and *75* have been widely studied (see Chiu *et al.*[11] and Sullivan and Bolton,[173] and references therein) as have methylated derivatives of *75*[174]; the selenium-containing analog [*76*] has also been recently reported.[11,124] The splittings are taken from Chiu *et al.*[11] which also describes the use of $AlCl_3$–CH_3NO_2 and CH_3CN–H_2SO_4 as oxidants (see also Sullivan and Bolton[173]). Molecular orbital

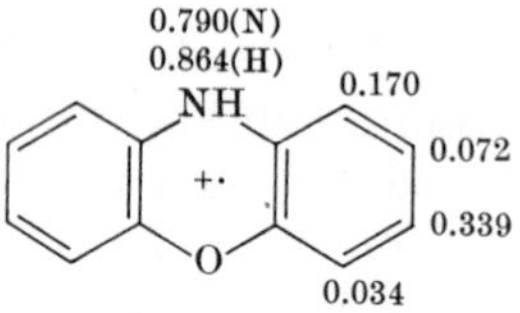

[*74*] 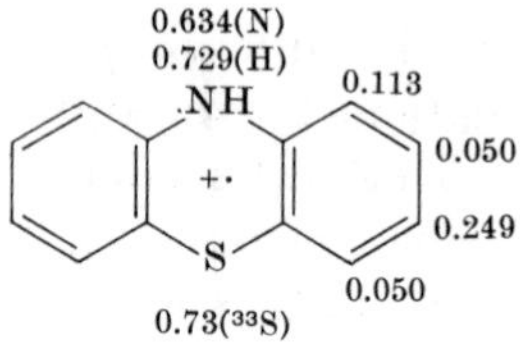   [*75*]

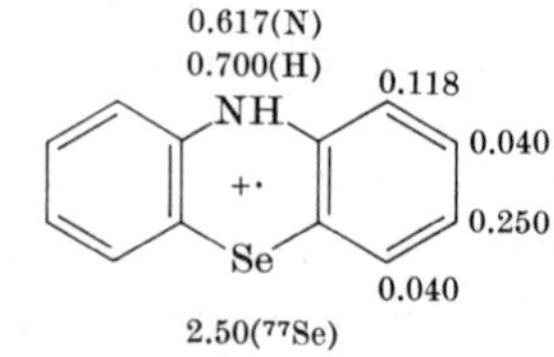

[*76*]

calculations for these radicals have been made using Hückel and McLachlan approaches[11,124,173]; typical parameters which lead to good agreement are[173] $h_S = 1.20$, $k_{CS} = 0.65$; $h_O = 1.70$, $k_{CS} = 0.80$, $h_N = 1.30$, $k_{CN} = 0.95$; and[124] $h_{Se} = 1.21$, $k_{C\,Se} = 0.57$. The radicals, unlike the parent molecules, appear to be planar. The spectrum of *76* exhibits increasing line broadening with increasing temperature, believed to be due to spin-rotational relaxation, which is expected to be marked for radicals with large g shifts.[11]

In a recent study,[175] controlled-potential electrolytic methods of radical generation were used in conjunction with ESR to investigate homogeneous electron-transfer reactions between some radicals of this

[173] P. D. Sullivan and J. R. Bolton, *J. Magn. Resonance* **1**, 356 (1969).

[174] J.-P. Billon, G. Cauquis, and J. Combrisson, *C.R. Acad. Sci. Paris* **253**, 1593 (1961).

[175] B. A. Kowert, L. Marcoux, and A. J. Bard, *J. Amer. Chem. Soc.* **94**, 5538 (1972).

type and their parent molecules. For example, for the cation radicals of phenoxazine, phenothiazine, and 10-methylphenothiazine at 23°C in CH_3CN with ClO_4^- as the counter-ion, the following second-order rate constants were determined: (4.5 ± 0.3), (6.7 ± 0.4), and $(2.2 \pm 0.3) \times 10^9$ liters mole^{-1} sec^{-1}, respectively. Correlation of the rates with electronic structure was explored.

A considerable number of further ESR investigations of cation radicals from phenothiazine and its derivatives, some of which have physiological activity, has been reported. For example, intermediacy of the cation radicals of phenothiazine and 3-hydroxyphenothiazine in the reactions of phenothiazine-5-oxide with sulfuric acid has been demonstrated, and mechanisms of reaction have been proposed.[176] The analogous reactions of phenoxathiin-5-oxide in H_2SO_4 to give the cation radicals of phenoxathiin itself [Eq. (13)] and of 3-hydroxyphenoxathiin have also been monitored.[177] In these investigations, electronic absorption spectra of the radicals and other intermediates were recorded.

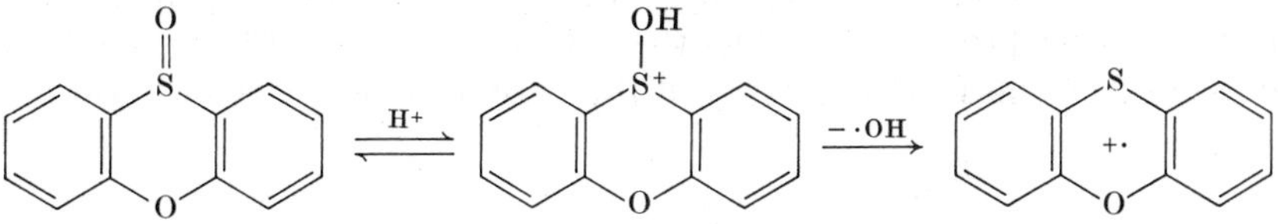

(13)

A variety of substituted phenothiazine drugs (alimepromazine, aminopromazine, fluphenazine, levomepromazine, perphenazine, and torecan) yield cation radicals in concentrated sulfuric acid solution,[178] and the cation radicals of the pharmacologically active promazine and chlorpromazine [77] have been prepared in several ways,[179] including

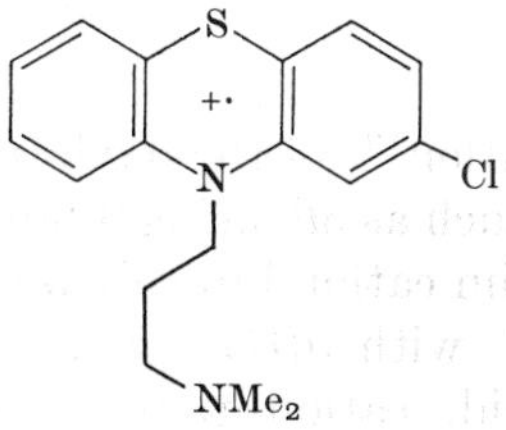

[77]

---

[176] H. J. Shine and E. E. Mach, *J. Org. Chem.* **30**, 2130 (1965); see also M. Răileanu, I. Radulian, O. Duliu, and S. Radu, *Collect. Czech. Chem. Commun.* **37**, 478 (1972).
[177] H. J. Shine and R. J. Small, *J. Org. Chem.* **30**, 2140 (1965).
[178] P. Machmer, *Z. Naturforsch. B* **21**, 934 (1966).
[179] H. Fenner and H. Möckel, *Tetrahedron Lett.* 2815 (1969).

enzymatic oxidation with peroxidase–H_2O_2 for the latter.[180] The psychotropic activity of this drug may be related to the stability of the associated radical.[180] Borg has reviewed the implications of ESR results regarding the possible physiological and pharmacological importance of radicals derived not only from some phenothiazine derivatives, including chlorpromazine and imipramine, but also from some alkaloids and several carcinogenic quinoline-based radicals (Chapter 7 of Swartz et al.[10]). In another investigation,[181] the interaction between the cation radical of chlorpromazine and deoxyribonucleic acid has been studied using ESR; evidence was adduced that the radical is intercalated in the DNA with its molecular plane perpendicular to the latter's helical axis. Depending on the alignment in the magnetic field of the capillary flow system employed, specific "oriented" spectra were observed.

Electron spin resonance has also been used in a novel fashion to help detect a sigmatropic rearrangement in the triethyl phosphite-induced conversion of [4-^{2}H]phenyl 2-nitrophenyl sulfide [78] into [3-^{2}H]-phenothiazine [79].[182] The identity of the product was verified by inspection of the ESR spectrum of the cation radical derived from it.

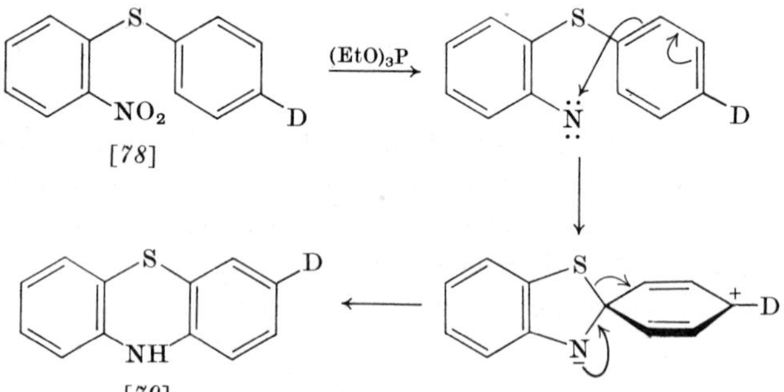

During an investigation[183] of the oxidative solvolysis of some metal bis- and trisdithienes, such as 80, using dilute sulfuric acid in anhydrous nitromethane, 1,4-dithiin cation radicals were detected using ESR. For example, 80 yields 81, with $a(H) = 0.206$ mT and $g = 2.006$. Using more concentrated acid, cation radicals from free cis-1,2-ethylene-dithiols were observed whose spectra were replaced slowly on standing by those of the 1,4-dithiin species. The open chain radicals were also

[180] L. H. Piette, G. Bulow, and I. Yamazaki, Biochim. Biophys. Acta 88, 120 (1964).
[181] S. Ohnishi and H. M. McConnell, J. Amer. Chem. Soc. 87, 2293 (1965).
[182] J. I. G. Cadogan, S. Kulik, and C. Thomson, J. Chem. Soc. D 436 (1970).
[183] G. N. Schrauzer and H. N. Rabinowitz, J. Amer. Chem. Soc. 92, 5769 (1970).

[80]   [81]   [82]

found to react with added acetone, generating radicals such as *82*, which
has $a(H) = 0.575$ mT and $g = 2.009$. Treatment[184] of biacetyl with
sulfuric acid and sodium dithionite, reagents which generate hydrogen
sulfide, produces a radical whose ESR spectrum shows a splitting of
0.217 mT (six equivalent hydrogen atoms); this species was assigned
the structure *83* ($R = CH_3$). Interestingly, when bis(trifluoromethyl)-
1,2-dithiete was dissolved in sulfuric acid it was oxidized directly to a
cation radical (*83*, $R = CF_3$) with $a(F) = 0.135$ mT. The unusually

[83]   [84]   [85]

stable bis-1,3-dithiolium cation radical [*84*] results from treatment of the
parent tetrasulfur-substituted olefin with Cl_2 in CCl_4[185]; it has $a(H) =$
0.126 mT and $g = 2.00838$. The 4,4'-bithiopyrylium cation radical
results from reduction of the corresponding dication with zinc in
acetonitrile at room temperature (cf. the reduction of nitrogen analogs
discussed below)[186]; MO calculations (including allowance for 3*d*
orbital participation by sulfur) have been reported for this radical and
also for the cation radical of thieno[3,2-*b*]thiophene [*85*],[187] which was
prepared from the parent with either $AlCl_3$–CH_3NO_2 at $-20°C$ or
$SbCl_5$–CH_2Cl_2 at $-60°C$. That the splittings are smaller than those for
the corresponding anion (see p.123) can be rationalized in terms of MO
calculations which allow *p* rather than *d* orbital participation.

One-electron reduction of the diquaternary cations of some suitable
nitrogen-containing heteroatomic compounds can yield stable cation
radicals. For example, electrolytic reduction in DMF in the presence of
perchloric acid or reduction with zinc dust in ethanol containing concentrated hydrochloric acid, of pyrazine, 4,4'-bipyridyl, and phenazine,

[184] G. A. Russell, R. Tanikaga, and E. R. Talaty, *J. Amer. Chem. Soc.* **94**, 6125 (1972).
[185] F. Wudl, G. M. Smith, and E. J. Hufnagel, *J. Chem. Soc. D* 1453 (1970).
[186] Z. Yoshida, T. Sugimoto, and S. Yoneda, *J. Chem. Soc. Chem. Commun.* 60 (1972).
[187] L. Lunazzi, G. Placucci, and M. Tiecco, *Tetrahedron Lett.* 3847 (1972).

leads to the cation radicals *86–88*, respectively.[188] The reductive method has also been used for some N-methylated species, and photolysis of bis-quaternary salts in ethanol has also been employed.[189]

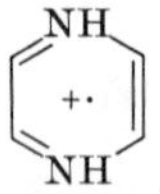

[*86*] 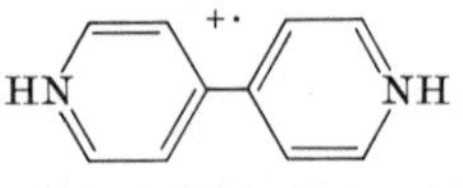   [*87*] 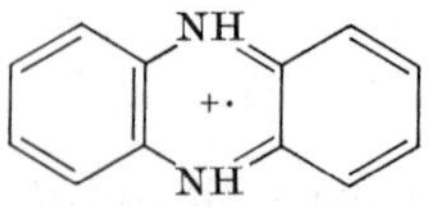   [*88*]

Electron spin resonance results indicate that spin density on carbon atoms adjacent to the nitrogen atoms in these radicals makes only a small contribution to the magnitude of $a(N)$; good results were obtained using simple Hückel calculations.[188]

The N,N'-dimethyl derivative of radical *87* is particularly stable and it has been the subject of considerable study. This "viologen" radical, the one-electron reduction product from the herbicide "paraquat," has been reported recently also to result, rather unusually, from di(4-pyridyl)ketone di- and monomethiodide in aqueous sodium hydroxide solution.[190] It also results[191] from the reduction of the 1-methyl-4-cyanopyridinium cation.

The importance of ESR studies to our understanding of the mechanism of action of some viologens as herbicides has been discussed (reference 10, Chapter 7). Recent results of relevance include the observation of spectra from the cation radicals of 1,2-bis(N-methyl-4-pyridyl)ethylene[192] and of some pyridinyl compounds with a methylene chain separating the aromatic rings (*89*, $n = 3, 4,$ and 5).[193] Electron spin resonance results indicate that for (*89*, $n = 3$ and 4) there is

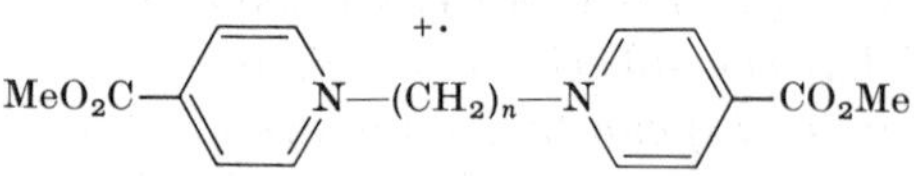

[*89*] 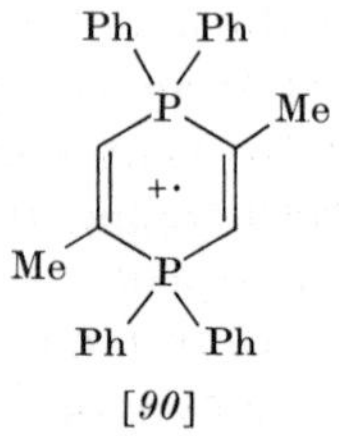   [*90*]

[188] B. L. Barton and G. K. Fraenkel, *J. Chem. Phys.* **41**, 1455 (1964).
[189] C. S. Johnson, Jr., and H. S. Gutowsky, *J. Chem. Phys.* **39**, 58 (1963); see also A. Castellano, J.-P. Catteau, and A. Lablache-Combier, *J. Chem. Soc. Chem. Commun.* 1207 (1972).
[190] F. E. Geiger, C. L. Trichilo, F. L. Minn, and N. Filipescu, *J. Org. Chem.* **36**, 357 (1971).
[191] E. M. Kosower and J. L. Cotter, *J. Amer. Chem. Soc.* **86**, 5524 (1964).
[192] J. W. Happ, J. A. Ferguson, and D. G. Whitten, *J. Org. Chem.* **37**, 1485 (1972).
[193] M. Itoh, *J. Amer. Chem. Soc.* **93**, 4750 (1971).

association between the opposite ends of the radical even at room temperature.

Electrochemical reduction of series of 1,4-diphosphoniacyclohexa-2,5-diene salts leads to ESR signals from the corresponding cation radicals, e.g., *90*[194]; analysis of the spectra suggests that the unpaired electron occupies a delocalized molecular orbital including all six ring atoms, and the unusually large variation of $a(^{31}P)$ with temperature leads to the suggestion that rapid conformational changes are taking place.

C. Neutral Radicals

Most of the heterocyclic neutral radicals investigated by ESR are of the long-lived variety, these usually being stabilized by delocalization of the unpaired electron either over a conjugated system or on to a nitroxide function. Exceptions have been the 2-furylmethyl radical[195] and the 2- and 3-thienylmethyl radicals[136,141] generated by continuous *in situ* photolysis of di-*t*-butyl peroxide in the presence of the corresponding methyl compounds (see p. 125). Restricted rotation and spin delocalization in these radicals have been discussed, and a *p* orbital model for sulfur in the thiophene species is preferred.[136] Another approach to the investigation of short-lived radicals is to prevent their further reaction by trapping within an inert matrix. For example, the 2-pyridyl [*91*] and 2-pyrimidyl [*92*] radicals have been prepared[196] by electron-transfer reactions of the appropriate bromides with sodium or potassium atoms in an inert matrix at 77°K, using a rotating cryostat. The splittings, particularly those for ^{13}C, establish that both *91* and *92* are σ radicals with the unpaired electron localized largely in an sp^2 hybrid orbital at the ring position where abstraction has occurred. Radical *91* and its 3- and 4-isomers have also been generated[197] in an argon matrix at 4°K using a similar method. *In situ* UV irradiation of

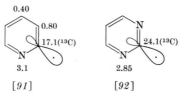

[*91*]   [*92*]

[194] R. D. Rieke, R. A. Copenhafer, A. M. Aguiar, M. S. Chattha, and J. C. Williams, Jr., *J. Chem. Soc. Chem. Commun.* 1130 (1972).

[195] L. D. Kispert, R. C. Quijano, and C. U. Pittman, Jr., *J. Org. Chem.* **36**, 3837 (1971).

[196] J. E. Bennett and B. Mile, *J. Phys. Chem.* **75**, 3432 (1971).

[197] P. H. Kasai and D. McLeod, Jr., *J. Amer. Chem. Soc.* **94**, 720 (1972).

the radicals leads to some interesting ring-opening reactions. Azabenzyl radicals have been obtained[198] by X irradiation of α-, β-, and γ-picoline in adamantane matrices where the spectra are isotropic because the radicals can undergo rapid reorientation (unlike the results described above[196]). Intermediate Neglect of Differential Overlap calculations on some of these radicals have been carried out.[197,198]

Stable neutral radicals containing sulfur whose stability and structure have been discussed include a series of trithienylmethyl and related radicals generated[199] by reduction of the corresponding perchlorate cations in DME with zinc powder. It appears that the 3-thienyl group is approximately as efficient as a phenyl ring at delocalizing the unpaired electron but that a 2-thienyl group is more effective than both. The 2,4,6-triphenylpyryl radical (*93*, X = O) has been prepared by zinc reduction of the perchlorate salt in cyclohexane at 25°C[200]; the observed splittings were assigned on the basis of spectra from deuterium-substituted analogs and both Hückel and McLachlan MO methods lead to good agreement with experiment for angles of out-of-plane twist of 28° for the 4-phenyl group and 42° for the 2- and 6-phenyl groups. For the analog (*93*, X = S), ^{33}S and ^{13}C splittings have been detected,[201] and calculations using both p and d models lead to similar angles of twist.

[*93*]   [*94*]

The 9-xanthyl radical (*94*, X = O, R = H) has been generated[202] by thermolysis of dixanthyl in hydrocarbon solvents between 180° and 260°C, and fairly good agreement with experiment results from Hückel and McLachlan calculations for a planar radical. For a range of alkyl-substituted radicals (*94*, X = O, R = Me, Et, *i*-Pr, *s*-Bu, and benzyl) the alkyl (β-hydrogen) splittings and their temperature dependences have been analyzed[203] in terms of the preferred conformations of substituents. Homolysis of 9,9'-diphenyldixanthyl yields *94* (X = O,

[198] R. V. Lloyd and D. E. Wood, *Mol. Phys.* **20**, 735 (1971).

[199] A. Mangini, G. F. Pedulli, and M. Tiecco, *J. Heterocycl. Chem.* **6**, 271 (1969).

[200] I. Degani, L. Lunazzi, and G. F. Pedulli, *Mol. Phys.* **14**, 217 (1968).

[201] I. Degani, L. Lunazzi, G. F. Pedulli, C. Vincenzi, and A. Mangini, *Mol. Phys.* **18**, 613 (1970).

[202] M. D. Sevilla and G. Vincow, *J. Phys. Chem.* **72**, 3635 (1968).

[203] M. D. Sevilla and G. Vincow, *J. Phys. Chem.* **72**, 3647 (1968).

R = Ph), for which calculations suggest an angle of twist for the 9-phenyl group of about 60°.[204] In another study of this type, the range of radicals examined was extended to include the sulfur[205–207] and selenium[206,207] analogs. Detailed analysis[205,206] of the spectra from some suitably deuterated xanthyl derivatives indicates that some earlier assignments[204,207] were partially incorrect; in particular, there is actually an inversion in the usually observed order of *ortho* and *para* hydrogen splittings for the 9-phenyl ring ($a_o > a_p$ in these examples), which can be accommodated within the MO framework for angles of twist of 70–80°. It was noted, however, that the MO methods do not give completely reliable results when applied to this sort of heterocyclic system. The symmetrical, planar triphenylmethyl-type radical, sesquixanthydryl, has been generated by thermolysis of the corresponding dimer.[208]

Radicals structurally related to those discussed above are the neutral radicals [*95*] and nitroxide radicals [*96*] derived from some heterocyclic analogs of anthracene. For example, treatment of phenoxazine and phenothiazine with dimethyl sulfoxide–acetic anhydride

[*95*]   [*96*]

leads[209] to ESR signals from *95* (X = O and S, respectively), as does oxidation with lead dioxide or lead tetraacetate.[210] The neutral radicals also result when the solutions of the respective cation radicals in CH_3CN are deprotonated with aqueous pH 7 buffer.[11] The radical *95* (X = Se) has now been prepared using PbO_2 in benzene.[124] Oxidation of the parent compounds with peroxides has been shown[124,210,211] for phenoxazine and phenothiazine to lead to neutral and nitroxide radicals, depending on the conditions, and oxidation of phenoselenazine with

[204] M. D. Sevilla and G. Vincow, *J. Phys. Chem.* **72**, 3641 (1968).
[205] L. Lunazzi, A. Mangini, G. Placucci, C. Vincenzi, and I. Degani, *Mol. Phys.* **19**, 543 (1970).
[206] L. Lunazzi, A. Mangini, G. Placucci, and C. Vincenzi, *J. Chem. Soc. Perkin Trans. 1* 2418 (1972).
[207] K. Maruyama, M. Yoshida, and K. Murakami, *Bull. Chem. Soc. Jap.* **43**, 152 (1970).
[208] E. Müller, A. Moosmayer, A. Rieker, and K. Scheffler, *Tetrahedron Lett.* 3877 (1967).
[209] Y. Tsujino, *Tetrahedron Lett.* 4111 (1968).
[210] C. Jackson and N. K. D. Patel, *Tetrahedron Lett.* 2255 (1967).
[211] K. Scheffler and H. B. Stegmann, *Tetrahedron Lett.* 3619 (1968).

t-butyl hydroperoxide or with *p*-nitroperbenzoic acid leads to ESR signals from the nitroxide (*96*, X = Se).[124] Photolysis of ethanolic solutions of the phenothiazine cation radical apparently[124,210] leads to the nitroxide radical (*96*, X = S). Molecular orbital calculations for *95* (X = O, S, Se) and *96* (X = O, S, Se) have been described[11,124] which reproduce satisfactorily the trends in the two series and which also reproduce "horizontal" trends (i.e., cation radical, neutral radical, and nitroxide for oxygen, sulfur, and selenium). Spin rotational line broadening and [77]Se splittings for *95* (X = Se) and *96* (X = Se) have been observed.[11,124] Particularly stable radicals reported include some dibenzo[*c,h*]phenothiazinyl neutral radicals,[212] some *t*-butyl derivatives of phenoxazinyl and phenoxazine nitroxide,[211] the tetraphenylpyrryl radical,[213] and the cation radical, neutral radical, and nitroxide from 1,3,6,8-tetra-*t*-butylcarbazole.[214]

Certain pyridinyl radicals, prepared by the reduction of quaternary pyridinium salts with zinc, potassium, or magnesium, are also stable in the absence of oxygen and other reactive species[215,216]; for example, the radical *97* has been purified by distillation *in vacuo*.[215] Intramolecular association in pyridinyl biradicals containing two such radical moieties joined by a methylene chain has been investigated using ESR[217] and stable 1,4-dihydro-1-silylpyridinyl radicals have also been reported.[218]

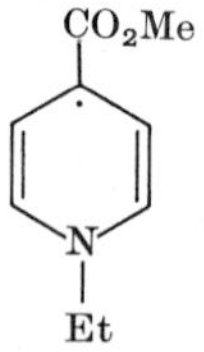

[*97*] 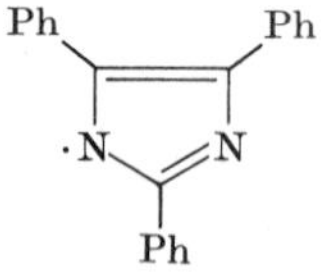   [*98*]

Stable neutral radicals containing more than one nitrogen atom in a ring include nitronyl nitroxides, such as those discussed earlier, ni-

[212] J. Brandt, G. Fauth, W. H. Franke, and M. Zander, *Chem. Ber.* **104**, 519 (1971); **105**, 1142 (1972).
[213] R. D. Allendoerfer and A. S. Pollock, *Mol. Phys.* **22**, 661 (1971).
[214] F. A. Neugebauer, H. Fischer, S. Bamberger, and H. O. Smith, *Chem. Ber.* **105**, 2694 (1972).
[215] E. M. Kosower and E. J. Poziomek, *J. Amer. Chem. Soc.* **86**, 5515 (1964).
[216] M. Itoh and S. Nagakura, *Bull. Chem. Soc. Jap.* **39**, 369 (1966).
[217] M. Itoh and E. M. Kosower, *J. Amer. Chem. Soc.* **90**, 1843 (1968).
[218] B. Schroeder, W. P. Neumann, J. Hollaender, and H.-P. Becker, *Angew. Chem. Int. Ed.* **11**, 850 (1972).

troxides from 1-hydroxyimidazoles and 1-hydroxyimidazole-3-oxides,[219] and the 2,4,5-triphenylimidazolyl radical [*98*] derived by photolysis of the corresponding 1,2'-dimer, the 1,1'-dimer,[220] and also from oxidation of 2,4,5-triphenylimidazole.[213,220] Photo-, thermo-, and piezochromic properties of the dimers depend on radical formation. Deuterium substitution was used as an aid to the assignment of splittings for *98*, and MO calculations indicate that, as suggested by the inspection of models, the 4- and 5-phenyl rings are twisted out of the plane of the imidazole ring by about 40°.[220]

Other stable radicals investigated using ESR include some verdazyls[221] (e.g., a series of *p*-substituted aryl derivatives [*99*] for which substituent effects were investigated[222]) and[223,224] tetrazolinyl radicals (e.g., *100*). It has been shown[224] that these radicals can be formed both

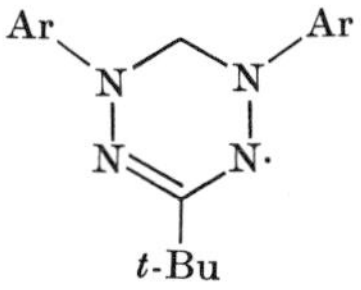

t-Bu t-Bu
[*99*] 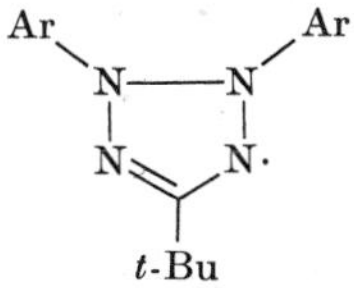   [*100*]

by the reduction of tetrazolium salts and by the one-electron oxidation of open-chain formazans.

IV. MISCELLANEOUS RADICALS

We conclude with a brief summary of further investigations involving the ESR spectra of heterocyclic radicals in which we refer the interested reader to the relevant reports for detailed discussion. In particular, the numerous applications of ESR to systems of biological importance are essentially beyond the scope of this review; however, an excellent comprehensive coverage of this topic has recently appeared.[10]

The reactions of 5-halouracils with hydroxyl radicals, produced in irradiated aqueous solution, have been monitored using ESR[225];

[219] K. Volkamer and H. W. Zimmermann, *Chem. Ber.* **103**, 296 (1970).
[220] N. Cyr, M. A. J. Wilks, and M. R. Willis, *J. Chem. Soc. B* 404 (1971); T. Hayashi, K. Maeda, and M. Morinaga, *Bull. Chem. Soc. Jap.* **37**, 1563 (1964); H. Ueda, *J. Phys. Chem.* **68**, 1304 (1964).
[221] P. Kopf, K. Morokuma, and R. Kreilick, *J. Chem. Phys.* **54**, 105 (1971).
[222] F. A. Neugebauer, *Tetrahedron* **26**, 4853 (1970).
[223] F. A. Neugebauer and G. A. Russell, *J. Org. Chem.* **33**, 2744 (1968); F. A. Neugebauer, *Chem. Ber.* **102**, 1339 (1969); *Tetrahedron* **26**, 4843 (1970).
[224] O. W. Maender and G. A. Russell, *J. Org. Chem.* **31**, 442 (1966).
[225] P. Neta, *J. Phys. Chem.* **76**, 2399 (1972).

spectra have also been reported[226] for the related radicals formed by photoreduction of alloxan, uracil, and some derivatives in aqueous solutions containing small amounts of a hydrogen donor (isopropanol). The reactions of $\cdot OH$ and $\cdot NH_2$ with some nucleosides[227] and pyrimidines[228] have also been studied using a flow system. In all these investigations, ESR allows unambiguous recognition of the delocalized radicals produced and hence the characterization of abstraction and addition reactions.

The radical *101* is observed during the autoxidation of ascorbic acid between pH 6.6 and 9.6.[229] It is interesting that hyperfine interaction

[*101*]

is only detected for three hydrogens: $a(1H) = 0.17$ and $a(2H) = 0.017$ mT. We believe that the larger splitting is from the β hydrogen, as expected, but that the other two apparently equivalent nuclei could be the γ hydrogen and one of the δ hydrogens; the two δ hydrogens may be rendered nonequivalent by the adjacent chiral γ carbon atom. The physiological importance of this species and of radicals from other vitamins and related compounds has been discussed.[10]

Considerable interest has been shown in radicals produced by one-electron oxidation[230] of a variety of metalloporphyrins (e.g., of Zn and Mg); in one case,[231] hyperfine interaction was detected from halide counter-ions (F^-, Cl^-, Br^-) present in solution with the cation radicals. The electronic structures of the oxidized metalloporphyrins have been discussed, as have those of some reduced porphins.[232] The assignment of signals detected during photosynthetic processes to the cation radical

[226] J. K. Dohrmann, R. Livingston, and H. Zeldes, *J. Amer. Chem. Soc.* **93**, 3343 (1971); J. K. Dohrmann and R. Livingston, *ibid.* **93**, 5363 (1971).

[227] H. Dertinger and C. Nicolau, *Biochim. Biophys. Acta* **199**, 316 (1970).

[228] C. Nicolau, M. McMillan, and R. O. C. Norman, *Biochim. Biophys. Acta* **174**, 413 (1969).

[229] C. Lagercrantz, *Acta Chem. Scand.* **18**, 562 (1964).

[230] J.-H. Fuhrhop and D. Mauzerall, *J. Amer. Chem. Soc.* **91**, 4174 (1969); J. Fajer, D. C. Borg, A. Forman, D. Dolphin, and R. H. Felton, *ibid.* **92**, 3451 (1970); A. Wolberg and J. Manassen, *ibid.* **92**, 2982 (1970).

[231] A. Forman, D. C. Borg, R. H. Felton, and J. Fajer, *J. Amer. Chem. Soc.* **93**, 2790 (1971).

[232] N. S. Hush and J. R. Rowlands, *J. Amer. Chem. Soc.* **89**, 2976 (1967).

of bacteriochlorophyll and to a plastochromanoxyl semiquinone radical has been discussed by Kohl (Swartz *et al.*,[10] Chapter 6). Participation by the anion radical of chlorophyll also appears to be implicated (see Swartz *et al.*,[10] Chapter 7), and it is clear that ESR has much to offer for the recognition of intermediates in the electron-transport chain and in the measurement of kinetics.

Another fruitful area has been the extended study of hyperfine splittings for some flavin free radicals and of the application of ESR to the detection of these intermediates in biologically important electron-transfer reactions. For example, the ESR spectrum of the lumiflavin anion radical [*102*] has been completely analyzed and the splittings assigned.[233] The corresponding mono- and diprotonated radicals have

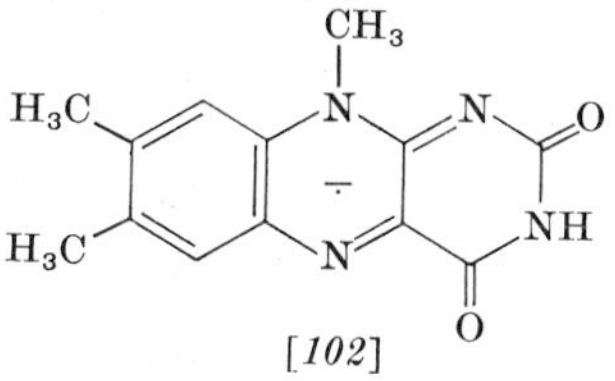

[*102*]

been detected using ESR,[234] and the protolytic equilibria involving these flavosemiquinones have been discussed.[235] Flavin radical chelates with Zn and Cd have been detected (^{67}Zn, ^{111}Cd, and ^{113}Cd splittings have been observed),[236] and a variety of alkyl derivatives[237] and sulfur-containing analogs[238] have been investigated. The interested reader is referred to a review by Beinert of the contribution of ESR to our understanding of the properties of free flavin radicals and also of the spectra and reactions of FMN, FAD, metal-free flavoproteins, and some metal flavoproteins of the xanthine variety (Swartz *et al.*,[10] Chapter 8).

V. APPENDIX

During 1973 a number of reports has appeared which update or extend work referred to in this review. Results on the following aspects are of particular relevance: conformations of radicals from five- and

---

[233] L. E. G. Eriksson and A. Ehrenberg, *Acta Chem. Scand.* **18**, 1437 (1964); *Arch. Biochem. Biophys.* **110**, 628 (1965).

[234] P. Hemmerich, C. Veeger, and H. C. S. Wood, *Angew. Chem. Int. Ed.* **4**, 671 (1965).

[235] A. Ehrenberg, F. Müller, and P. Hemmerich, *Eur. J. Biochem.* **2**, 286 (1967).

[236] A. Ehrenberg, L. E. G. Eriksson, and F. Müller, *Nature (London)* **212**, 503 (1966).

[237] W. H. Walker and A. Ehrenberg, *FEBS Lett.* **3**, 315 (1969).

[238] H. Fenner, *Tetrahedron Lett.* 617 (1970).

six-membered alicyclic compounds (including THF and tetrahydro-pyran),[239] and geometry of some oxygen-bridged 2-norbornyl radicals,[240] conformational interconversion in 4-alkylpiperidine nitroxides,[241] tri-alkylhydrazyl radicals,[242] the geometry of the 9-phenylacridine anion,[243] anion radicals from cyclazines,[244,245] thiophenes and fused thiophenes,[246,247] and from some 6a-thiathiophthenes,[248] conformation-al isomerism in cation radicals of some alkylthiothiophenes,[249] and de-tailed investigations of radical intermediates in the oxidation of ascorbic acid and related substances.[250] There has also been a review[251] of hydrazidinyl radicals including 1,2,4,5-tetraazapentenyls, verdazyls, and tetrazolinyls.

In addition, several more novel investigations have been described. Anion radicals have been generated from furan, isoxazole, and oxazole,[252] and from pyrrole, pyrazole, imidazole, and indole,[253] in argon matrices at 4°K by reaction of the parent heterocycles with sodium atoms. For the oxygen-containing compounds, the vinyl-type radicals detected arise via the rupture of a bond to oxygen; in contrast,[253] for pyrrole and related compounds the rings remain intact and anion radicals of tautomeric forms are detected. The reactions of radiolytically-produced radicals (e.g., hydroxyl) with furan and some derivatives in aqueous solution have been investigated[254]; while in acidic media the furan moiety remains intact, in basic solution ring-opening follows ·OH

[239] C. Corvaja, G. Giacometti, and M. Brustolon, Z. Phys. Chem. (Frankfurt) 82, 272 (1972).

[240] T. Kawamura, T. Koyama, and T. Yonezawa, J. Amer. Chem. Soc. 95, 3220 (1973).

[241] R. E. Rolfe, K. D. Sales, and J. H. P. Utley, J. Chem. Soc. Perkin Trans. 2 1171 (1973).

[242] S. F. Nelsen and R. T. Landis, II, J. Amer. Chem. Soc. 95, 6454 (1973).

[243] A. Lomax, L. S. Marcoux, and A. J. Bard, J. Phys. Chem. 76, 3958 (1972).

[244] F. Gerson, J. Jachimowicz, B. Kowert, and D. Leaver, Helv. Chim. Acta 56, 258 (1973).

[245] F. Gerson, J. Jachimowicz, and D. Leaver, J. Amer. Chem. Soc. 95, 6702 (1973).

[246] M. Guerra, G. F. Pedulli, and M. Tiecco, J. Chem. Soc. Perkin Trans. 2 903 (1973).

[247] G. F. Pedulli, M. Tiecco, A. Alberti, and G. Martelli, J. Chem. Soc. Perkin Trans. 2 1816 (1973).

[248] F. Gerson, J. Heinzer, and M. Stavaux, Helv. Chim. Acta 56, 1845 (1973).

[249] C. M. Camaggi, L. Lunazzi, and G. Placucci, J. Chem. Soc. Perkin Trans. 2 1491 (1973).

[250] G. P. Laroff, R. W. Fessenden, and R. H. Schuler, J. Amer. Chem. Soc. 94, 9062 (1972).

[251] F. A. Neugebauer, Angew. Chem. Int. Ed. 12, 455 (1973).

[252] P. H. Kasai and D. McLeod, Jr., J. Amer. Chem. Soc. 95, 4801 (1973).

[253] P. H. Kasai and D. McLeod, Jr., J. Amer. Chem. Soc. 95, 27 (1973).

[254] R. H. Schuler, G. P. Laroff, and R. W. Fessenden, J. Phys. Chem. 77, 456 (1973).

addition. Related investigations[255] of pyrroles and imidazoles indicate that there is no ring-opening at high pH. Hydroxyl radicals produced similarly are found[256] to react with 5-nitrouracil and 5-nitro-2-furoic acid to give ESR spectra from radicals formed via oxidative denitration. The spin trapping of radicals formed in the photochemical reaction between hydrogen peroxide and some pyrimidine bases, nucleosides, nucleotides, and yeast nucleic acid has also been reported.[257] The reaction between the hydroxyisopropyl radical ($\cdot CMe_2OH$) and some nitrogen-containing heteroaromatic compounds has been used[258] to generate radicals from pyridine and some derivatives, pyrazine, pyrimidine, and pyridazine; electron transfer from the donor radical is followed by protonation. Similarly,[259] pyridazine and pyrazine are susceptible to reduction by radiolytically-prepared solvated electrons; the initially-formed anions also undergo rapid protonation at nitrogen to form pyridinyl and pyrazinyl radicals.

Recently, evidence has been adduced[260] for the formation of dimer cation radicals during the one-electron oxidation (e.g., with $\cdot OH$) of some sulfides, including tetrahydrothiophene, and in several studies of phosphoranyl radicals (including some spiro derivatives) structural, mechanistic, and kinetic aspects have been investigated.[261–265] Other interesting reports involving heterocyclic radicals include the kinetic investigation[266] of the reactions of alkylaminyl radicals (e.g., 2,2,6,6-tetramethylpiperidyl) and the detection of radical intermediates in the decomposition[267] of N,N'-diethyoxycarbonyl-4,4',N,N'-tetrahydro-4, 4'-bipyridyl and in the reactions[268] of acylpyridinium salts with alkali.

[255] A. Samuni and P. Neta, *J. Phys. Chem.* **77**, 1629 (1973).

[256] P. Neta and C. L. Greenstock, *J. Chem. Soc. Chem. Commun.* 309 (1973).

[257] C. Lagercrantz, *J. Amer. Chem. Soc.* **95**, 220 (1973).

[258] H. Zeldes and R. Livingston, *J. Phys. Chem.* **76**, 3348 (1972); **77**, 2076 (1973).

[259] R. W. Fessenden and P. Neta, *Chem. Phys. Lett.* **18**, 14 (1973).

[260] B. C. Gilbert, D. K. C. Hodgeman, and R. O. C. Norman, *J. Chem. Soc. Perkin Trans.* 2 1748 (1973).

[261] D. Griller and B. P. Roberts, *J. Chem. Soc. Perkin Trans.* 2 1339, 1416 (1973); *J. Organometal. Chem.* **42**, C47 (1972).

[262] R. W. Dennis and B. P. Roberts, *J. Organometal. Chem.* **47**, C8 (1973).

[263] A. G. Davies, R. W. Dennis, D. Griller, K. U. Ingold, and B. P. Roberts, *Mol. Phys.* **25**, 989 (1973).

[264] A. G. Davies, D. Griller, and B. P. Roberts, *J. Chem. Soc. Perkin Trans.* 2 2224 (1972).

[265] G. B. Watts, D. Griller, and K. U. Ingold, *J. Amer. Chem. Soc.* **94**, 8784 (1972).

[266] J. R. Roberts and K. U. Ingold, *J. Amer. Chem. Soc.* **95**, 3228 (1973).

[267] P. Atlani, J. F. Biellmann, R. Brière, and A. Rassat, *Tetrahedron* **28**, 5805 (1972).

[268] M. Frangopol, P. T. Frangopol, C. L. Trichilo, F. E. Geiger, and N. Filipescu, *J. Org. Chem.* **38**, 2355 (1973).

·4·

Fluorescence and Phosphorescence of Heterocyclic Molecules

STEPHEN G. SCHULMAN

COLLEGE OF PHARMACY, UNIVERSITY OF FLORIDA
GAINESVILLE, FLORIDA 32601

I. INTRODUCTION

This chapter will consider the emission of light by individual molecules in solution, especially heterocyclic molecules, which are electronically excited as a result of the absorption of visible or ultraviolet light. This phenomenon is termed photoluminescence and encompasses fluorescence and phosphorescence spectroscopy. To date, the luminescences of only those heterocycles containing nitrogen, oxygen, and sulfur as heteroatoms have been appreciably studied and it is to these molecules that attention will be directed. The luminescences of molecules induced by chemical reactions (chemiluminescence), excitation by electric fields (electroluminescence), and by thermal excitation (thermoluminescence), as well as the luminescences of crystalline organic and inorganic substances, are beyond the scope of this work and the reader is referred to several excellent reviews of these subjects in the literature.[1,2]

The photoluminescences of organic molecules have been employed predominately in quantitative analysis. However, careful evaluation of fluorescence and phosphorescence spectra often permits qualitative identification of molecular species at concentrations far lower ($\sim 1 \times 10^{-9}$ M) than those normally employed with the more widely employed spectroscopic techniques (e.g., IR and NMR). In addition, fluorescence and phosphorescence measurements often yield valuable information about the electronic structures of electronically excited molecules and about processes occurring in electronically excited states. These techniques are therefore of prime importance to the photochemist.

The physical and chemical factors governing the absorption of light by heterocyclic molecules and conjugated molecules in general have already been discussed at length in previous volumes in this series[3,4]. This chapter will therefore be primarily concerned with the physical and chemical processes following the transitions of π and nonbonded electrons caused by the absorption of visible and near ultraviolet light, which lead to and compete with fluorescence and phosphorescence, as well as the luminescence phenomena themselves.

Luminescence processes in molecules are affected by molecular structure, solvent, and chemical processes occurring in electronically excited states. The understanding of the interplay of all these factors on molecular luminescence can be greatly simplied by first dealing with the fundamental physical processes which govern photoluminescence and those aspects of luminescence spectroscopy which are derived from the electronic and geometrical structures of the isolated molecules. In principle, the latter subjects entail the study of molecules in the gas phase. However, gas-phase luminescence experiments are often difficult to arrange, and to a very good approximation the spectral properties of isolated molecules can be conveniently studied in dilute solutions in nonpolar nonhydrogen bonding solvents (e.g., hydrocarbon solvents).

[1] H. Kallman and G. Marmor-Spruch (eds.), "Luminescence of Organic and Inorganic Materials." Wiley, New York, 1962.

[2] R. M. Hochstrasser, *Rev. Mod. Phys.* **34**, 531 (1962).

[3] S. F. Mason, *in* "Physical Methods in Heterocyclic Chemistry" (A. R. Katritzky, ed.), Vol. 2, p. 1. Academic Press, New York, 1963.

[4] W. Armarego, *in* "Physical Methods in Heterocyclic Chemistry" (A. R. Katritzky, ed.), Vol. 3, p. 67. Academic Press, New York, 1971.

II. PHOTOPHYSICAL PROCESSES IN ISOLATED ELECTRONICALLY EXCITED MOLECULES[5,6]

Excitation of a molecule from its ground electronic state to one of its excited electronic states of the same multiplicity (an excited singlet state for most organic molecules) often results in the production of a vibrationally excited state within the excited electronic state. The natural tendency of the vibrationally excited molecule is to dissipate its excess vibrational energy and thereby arrive at the lowest vibrational level belonging to the excited electronic state which is compatible with the ambient temperature. For simplicity, vibrational relaxation of this type will be assumed to proceed to the lowest vibrational level (vibrational quantum number $V = 0$) of the electronically excited state, regardless of the solution temperature.

The process of vibrational relaxation within the excited state is very rapid, requiring less than 10^{-13} sec for completion. Because the vibrational levels of any electronic state are quantized (i.e., discrete), the loss of vibrational energy must proceed in a stepwise fashion. The excess vibrational energy is dissipated in collisions with molecules in the environment (the solvent), each collision removing a quantum of vibrational energy. The less vibrationally excited the electronically excited molecule is, the fewer collisions are required to produce relaxation to the lowest vibrational level of the excited electronic state.

Once an excited molecule has relaxed to the lowest vibrational level of the electronically excited state, it can lose excitation energy only by going to a lower electronic energy level (i.e., the optical electron drops to a lower energy orbital). This can be accomplished in several ways. If the higher vibrational levels of the lower electronic state overlap the lower vibrational levels of the higher electronic state (that is, if the nuclear configurations and energies of the two electronic states are identical during a low energy vibration of the upper electronic state and a high energy vibration of the lower electronic state) the upper and lower electronic states will be in a transient thermal equilibrium which will permit population of the lower electronic state. This is known as internal conversion. Vibrational relaxation of the lower electronic state then follows as before.

If the lower vibrational levels of the upper electronic state do not overlap the higher vibrational levels of the lower electronic state but

[5] N. Turro, "Molecular Photochemistry," Chapters 1–4. Wiley, New York, 1972.
[6] J. D. Winefordner, S. G. Schulman, and T. C. O'Haver, "Luminescence Spectrometry in Analytical Chemistry." Wiley, New York, 1972.

are separated by a small gap (a few vibrational quanta wide), internal conversion may still take place by quantum mechanical "tunneling." The probability of tunneling decreases as the difference in energy between the lower vibrational levels of the upper electronic state and the upper vibrational levels of the lower electronic state increases.

If the energy separation of the upper and lower electronic states is great enough that direct vibrational coupling is impossible and tunneling is improbable, another process may be responsible for deactivation of the upper electronic state. This process carries the excited molecule, in one step, through the large gap between the lowest vibrational level of the upper electronic state to any one of a number of vibrational levels of the lower electronic state. The excess energy in this case is released as a photon of visible or ultraviolet light whose frequency depends on the difference in energy between the lowest vibrational level of the upper electronic state and the vibrational level of the lower electronic state to which the radiative transition occurs. Following radiative transition, the molecule undergoes vibrational relaxation to the lowest vibrational level of the lower electronic state. The radiative transition from the upper to the lower excited state of the same multiplicity is termed fluorescence.

If a molecule has a large number of modes of vibration in the lower state, overlap between the upper and lower state is probable, and the lower vibrational modes of the higher electronic state will be able to excite higher vibrational modes of the lower electronic state with high probability. Hence, vibrational dissipation of excitation energy and thus internal conversion will be favored. It is for this reason that aliphatic molecules and others which do not have rigid molecular skeletons and thus have many vibrational degrees of freedom rarely exhibit fluorescence. The aromatic molecules with their rigid ring structures are the class of compounds which most frequently show fluorescence.

In most aromatic molecules, the excited electronic states are much closer together than are the ground state and the lowest excited state. Consequently, efficient internal conversion almost always precludes the possibility of fluorescence arising as the result of electronic transition between states other than the ground state and the lowest excited state of the same multiplicity as the ground state.

The intrinsic lifetime of the lowest excited singlet state of an aromatic molecule is of the order of 10^{-8} sec. Consequently, even if a molecule cannot pass efficiently from its lowest excited singlet state to the ground state by internal conversion, other rapid processes occurring during the lifetime of the lowest excited singlet state may compete with fluorescence.

One of the more important processes to compete with fluorescence for deactivation of the lowest excited states of aromatic molecules is intersystem crossing (change of multiplicity) from the lowest excited singlet to the lowest triplet state. Intersystem crossing entails a change in spin angular momentum which, in isolated molecules, is a forbidden process because classically it violates the law of conservation of angular momentum. However, in virtually all molecules there is always some degree of coupling of spin and orbital electronic motions because quantum mechanically there is a finite probability of finding the electrons and hence the electronic spin vectors as well as the orbital angular momentum vectors at the atomic nuclei of the molecule—a condition which favors spin–orbital coupling. The coupling of spin and orbital angular momenta results to some degree in the loss of the concept of the molecular electronic spin as a well-defined constant of motion of the molecule (i.e., molecular spin is not a good molecular quantum number). As a result there is a finite probability that for any spatial molecular electronic configuration, except that of the ground singlet (in which all occupied orbitals are doubly occupied and the Pauli exclusion principle forbids the existence of a state with two unpaired electrons), the molecule with an even number of electrons may be in either a singlet or a triplet state. Thus, the lowest triplet state may be radiationlessly populated from the lowest excited singlet state, the efficiency of intersystem crossing depending on the details of molecular structure which favor or inhibit spin–orbital coupling. As we shall soon see, the presence of heteroatoms in an aromatic ring strongly favors intersystem crossing.

Following intersystem crossing, the molecules populating the lowest triplet state are rapidly thermalized by the usual vibrational relaxation process so that all phenomena arising from the lowest triplet state will be assumed to originate from the lowest vibrational level of that electronic state.

Molecules in the lowest excited triplet state can return to the ground state by either a radiative or a nonradiative transition. Because a change in multiplicity is involved in the transition from excited triplet to ground singlet state, the transition is forbidden and will occur only with low probability. Spin-forbidden transitions of either the radiative or nonradiative type occur with 10^{-3}–10^{-9} of the rates of their spin-allowed counterparts. Because the probabilities of spin-forbidden transitions are low, their inverses, the lifetimes of triplet states, are high, lasting from 10^{-6} to several seconds.

Nonradiative deactivation of the triplet state is termed triplet–singlet intersystem crossing and is favored by a small separation between the lowest triplet and ground singlet states.

The radiative transition from an upper electronic state to a lower electronic state of different multiplicity is called phosphorescence. Phosphorescence, like fluorescence, is most often observed in molecules having rigid molecular skeletons and large energy separations between the lowest triplet state and the ground singlet state (i.e., aromatic molecules). Because triplet states are so long lived, chemical and physical processes in solution, as well as intersystem crossing, compete effectively with phosphorescence for depopulation of the lowest excited triplet state. This is the primary reason for the importance of the triplet state in photochemistry. Except for the shortest-lived phosphorescences, collisional deactivation by solvent molecules, quenching by paramagnetic species (e.g., oxygen), photochemical reactions, and certain other processes preclude the observation of phosphorescence in fluid media. Rather, phosphorescence is normally studied in glasses at liquid nitrogen temperature or in solution in very viscous liquids where collisional processes cannot completely deactivate the triplet state. The sequence of events leading to fluorescence and phosphorescence is depicted in Fig. 1.

Because fluorescence normally originates only from the lowest excited singlet state and phosphorescence only from the lowest triplet state, only one fluorescence band and one phosphorescence band may be observed from any given molecular species. This is in contrast to electronic absorption, in which several absorption bands may be observed in the spectrum of a single molecular species. Since phosphorescence is precluded in fluid solutions, only a single emission band, that due to fluorescence, may be observed from a single molecular species. However, at low temperatures in rigid matrices both fluorescence and phosphorescence may appear in the emission spectrum if both types of emission are comparable in intensity so that one does not completely mask the other. In this case the band occurring at longer wavelength will be the phosphorescence band because the lowest triplet state always lies lower in energy than the lowest excited singlet state. This is a consequence of the lower repulsive energy in the triplet state than in the singlet state of the same spatial configuration which arises from the greater average separation of the electrons in the triplet state demanded by the Pauli exclusion principle. In low temperature luminescence spectroscopic measurements, phosphorescence is usually dis-

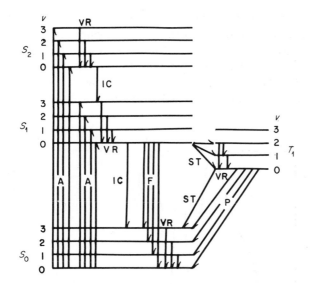

Fig. 1. Photophysical processes of conjugated molecules. Electronic absorption (A) from the lowest vibrational level ($v = 0$) of the ground state (S_0) to the various vibrational levels ($v = 0, 1, 2, 3$) of the excited singlet states (S_1 and S_2) is followed by rapid, radiationless internal conversion (IC) and vibrational relaxation (VR) to the lowest vibrational level ($v = 0$) of S_1. Competing for deactivation of the lowest excited singlet state S_1 are the radiationless internal conversion and singlet–triplet intersystem crossing (ST) as well as fluorescence (F). Fluorescence is followed by vibrational relaxation (VR) in the ground state. Intersystem crossing (ST) is followed by vibrational relaxation (VR) in the triplet state (T_1). Phosphorescence (P) and nonradiative triplet–singlet intersystem crossing (ST) return the molecule from the triplet state (T_1) to the ground state (S_0). Vibrational relaxation in S_0 then thermalizes the "hot" ground state molecule.

tinguished from fluorescence by taking advantage of the differences in the mean lifetimes of the two processes. For very long-lived phosphorescences (> 0.1 sec), mechanical chopping of the exciting light eliminates the fluorescence signal while the long-lived phosphorescence persists as an afterglow. For short-lived phosphorescences (< 0.1 sec), electronic chopping techniques are employed to distinguish between fluorescence and phosphorescence.

In an electronic absorption spectrum, the longest wavelength transition takes place between the ground and lowest excited singlet states, the same states involved in the fluorescence transition. If the spacing between vibrational sublevels in ground and lowest excited singlet states are equal in a given molecular species, the longest wavelength

absorption band and the fluorescence band will appear as mirror images
of one another when plotted on an abscissa linear in frequency or energy,
with the longest wavelength vibrational feature of the absorption spectrum and the shortest wavelength vibrational feature of the fluorescence
spectrum (the O–O bands) ideally coinciding. This is illustrated in
Fig. 2. In aromatic molecules with no exocylic functional groups the

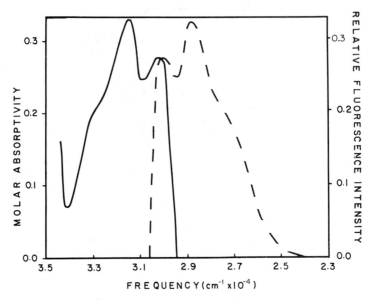

Fig. 2. Illustrating the mirror image relationship between the long wavelength
absorption (———) and fluorescence (– – –) bands of carbazole. Both spectra were taken
in cyclohexane. The fluorescence spectrum was adjusted in size to fit the scale of the
absorption spectrum. Note the near coincidence of the lowest frequency vibrational
feature of the absorption and the highest frequency vibrational feature of the fluorescence
(the O–O bands of absorption and fluorescence).

mirror image relationship between absorption and fluorescence is
usually observed in nonpolar solvents. However, when exocyclic
functional groups are attached to the aromatic ring, the geometry of
the exocyclic group and hence the vibrational structure of the molecule
is often different in the excited state from that in the ground state with
resulting loss of the perfect mirror image relationship between absorption and fluorescence.

III. THE EFFECT OF MOLECULAR STRUCTURE ON FLUORESCENCE AND PHOSPHORESCENCE SPECTRA

The most characteristic features of fluorescence and phosphorescence spectra are their positions in the electromagnetic spectrum and their intensities. These properties are functions of molecular structure and are of prime concern to the organic chemist. In this section the relationships between intrinsic molecular structure and luminescence spectral properties will be examined.

A. General Considerations

Fluorescence and phosphorescence spectral bands arise from electronic transitions between the lowest excited singlet or triplet state and the ground state of a given molecule. The position of the luminescence band in the electromagnetic spectrum, corresponding to a fluorescence or phosphorescence transition, depends on the energy separation between the excited singlet or triplet state and the ground electronic state and is given by the Planck frequency relation,

$$E_s - E_0 = h\nu_f = h(c/\lambda_f) \tag{1}$$

or

$$E_T - E_0 = h\nu_p = h(c/\lambda_p) \tag{2}$$

where E_s, E_T and E_0 are the energies of the lowest excited singlet state, the lowest triplet state, and the ground singlet state, respectively, h is Planck's constant, c is the velocity of light, ν_f and ν_p are the frequencies, and λ_f and λ_p are the wavelengths of fluorescent and phosphorescent light emitted, respectively. According to Eqs. (1) and (2) the greater the separation between E_s and E_0 or between E_T and E_0, the greater will be the frequencies and the shorter will be the wavelengths of fluorescence and phosphorescence. In the molecular orbital theory, $E_s - E_0$ and $E_T - E_0$ depend on the difference in energy between the highest occupied and lowest unoccupied molecular orbitals in the ground state of the molecule (which is the same for $E_s - E_0$ and $E_T - E_0$) and the differences in interelectronic repulsion energies for the electronic configurations corresponding to E_s, E_T, and E_0. Because the interelectronic repulsion energy is lower in the triplet state than in the lowest excited singlet state, phosphorescence occurs at longer wavelengths than fluorescence.

It is beyond the scope of this work to go into details of the molecular orbital theory of conjugated molecules. This subject is treated in

Mason,[3] Widefordner *et al.*,[6] and Murrell[7] and it will suffice to say that the electronic spectra of aromatic molecules can be, at least qualitatively, rationalized in terms of a particle (the optical electron) in a linear box (the aromatic ring system) model. In this model, the longer the box, the smaller is the energy separation between highest occupied and lowest unoccupied orbitals. Thus in the linearly annellated series, benzene ($\lambda_f = 262$ nm, $\lambda_p = 339$ nm), naphthalene ($\lambda_f = 314$ nm, $\lambda_p = 470$ nm), and anthracene ($\lambda_f = 379$ nm, $\lambda_p = 670$ nm), the separations between the highest occupied and lowest unoccupied orbitals decrease and the wavelengths of fluorescence and phosphorescence increase, with increasing extension of conjugation in the direction of linear annellation. However, in phenanthrene, which is tricyclic but angularly annellated, fluorescence ($\lambda_f = 354$ nm) is at longer wavelength than in naphthalene but at shorter wavelength than in anthracene because the linear extension of the conjugated system of naphthalene by angular annellation is not as great as that in anthracene. The phosphorescence of phenanthrene ($\lambda_p = 461$ nm) is slightly shorter in wavelength than that of naphthalene even though phenanthrene is more extensive. This is likely a result of the differences in the interelectronic repulsive energies between the ground and lowest triplet states of phenanthrene relative to those in the linear polyacenes and serves to underline the fact that spectral interpretations based solely on orbital considerations are only approximate.

Exocyclic substituents on aromatic rings affect the positions of the fluorescence and phosphorescence bands relative to those of the parent hydrocarbons in ways which depend on the natures of the electronic interactions between the substituents and the aromatic rings in ground and excited states. Groups which have lone electron pairs (e.g., $-NH_2$, $-OH$, $-SH$) which can be transferred into vacant π orbitals belonging to the aromatic ring are best treated by considering the lone pair (in the ground state of the molecule) as residing in a molecular orbital, largely localized on the exocyclic group, having an energy slightly lower than that of a pure atomic $2p$ orbital and considerably greater than that of the highest occupied π orbital of the aromatic ring (Fig. 3). Thus the energy gap between the highest occupied and lowest unoccupied orbitals of the substituted molecule is considerably less than that between the highest occupied and lowest unoccupied orbitals of the unsubstituted (parent) molecule. Absorption, fluorescence, and phosphorescence of the

[7] J. N. Murrell, "Theory of the Electronic Spectra of Organic Molecules." Methuen, London, 1963.

substituted molecule therefore occur at longer wavelengths than in the unsubstituted molecules. For example, while naphthalene fluoresces at 314 nm and phosphoresces at 470 nm, 1-naphthylamine fluoresces at 372 nm and phosphoresces at 527 nm.

Groups having localized, vacant (antibonding), low energy π orbitals

$$\text{(e.g., } -\overset{\displaystyle O}{\overset{\|}{C}}-OH, \; -\overset{\displaystyle O}{\overset{\|}{C}}-H, \; -C{\equiv}N)$$ in the ground state of the substituted molecule introduce a vacant orbital between the highest occupied and lowest unoccupied π orbitals of the unsubstituted molecule (Fig. 3). The energy gap between the highest occupied and the lowest unoccupied orbitals of the substituted molecule is thus smaller than that between

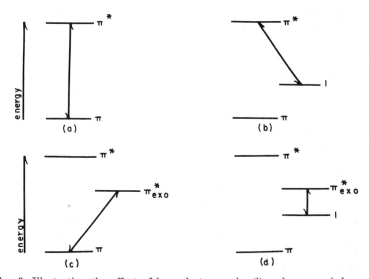

Fig. 3. Illustrating the effect of lone electron pairs (l) and unoccupied π orbitals belonging to exocyclic substituents (π^*_{exo}) on the energies of the electronic transitions of an aromatic molecule which give rise to the longest wavelength absorption and emission spectra. (a) In the unsubstituted hydrocarbon, the optical spectra of interest arise from transition between the highest occupied orbital (π) and the lowest unoccupied orbital (π^*). (b) Introduction of a heteroatom or exocyclic group having a lone or non-bonded electron pair results in the promotion or demotion of an electron between the lone or nonbonded pair orbital and the π^* orbital of the ring giving rise to the lowest energy optical transitions. (c) Introduction of an exocyclic group having a low-lying vacant π orbital results in transition between the π orbital of the ring and the π^*_{exo} orbital of the exocyclic group being lowest in energy. (d) Introduction of both an electron donor (l) and electron acceptor (π^*_{exo}) orbital results in the auxochromic effect, with transition between the latter two orbitals lying lowest in energy.

the highest occupied and lowest unoccupied orbitals of the unsubstituted molecule. Fluorescence and phosphorescence thus appear at longer wavelengths in the substituted molecule than in the unsubstituted molecule. For example, the fluorescence of 2-naphthoic acid lies at 344 nm in hexane and the phosphorescence of this compound at 488 nm. Both emissions are at longer wavelengths than their counterparts in naphthalene.

In the electron donating groups, absorption entails promotion of an electron from a lone pair to a vacant π orbital of the aromatic ring. Correspondingly, the emission processes involve transfer of electronic charge from the aromatic ring back to the exocyclic group. In the electron-withdrawing groups, absorption entails promotion of an electron from a π orbital of the aromatic ring to a vacant π orbital of the exocyclic group while fluorescence and phosphorescence entail transfer of charge from the exocyclic group back to the aromatic ring. These processes are known as intramolecular charge transfer transitions, the exocyclic groups in a sense being regarded as separate entities interacting with the aromatic ring. Owing to the smaller degree of interelectronic repulsion in the triplet state than in the lowest excited singlet state, the dipole moment changes accompanying transitions between the triplet and ground states are generally smaller than those between the lowest excited singlet and ground states. This is reflected in a smaller degree of intramolecular charge transfer in singlet–triplet transitions than in singlet–singlet transitions with the result that substituents in aromatic rings do not shift the phosphorescence as much as they do the fluorescence or longest wavelength absorption bands. Charge transfer accompanying excitation to the lowest excited singlet state usually results in the excited molecule having a greater dipole moment than the ground state molecule. Frequently, the exocyclic group is reoriented (rehybridized) from its ground state configuration to effect more efficient conjugation in the excited state. Rehybridization in the excited state involves nuclear motions and therefore a vibrational relaxation process and is not reflected in the absorption spectrum. After fluorescence, rehybridization accompanied by vibrational relaxation occurs in the ground state to give the "normal" ground state molecule and this process is not reflected in the fluorescence spectrum. If the degree of interaction between the exocyclic group and the aromatic ring is greater in the excited state than in the ground state (as is usual), the energy of vibrational relaxation in the excited state will be greater than that in the ground state. In this case the shifting effect of the substituent on the fluorescence relative to that of the unsubstituted molecule will be greater than the shifting effect of the substituent on the absorption spectrum (i.e., the fluorescence of the substituted molecule will occur farther downfield from that of the unsubstituted molecule than will the absorption spectrum of the substituted molecule from that of the unsubstituted molecule). For example, the absorption spectrum of 9-anthroic acid is very much like that of anthracene, but the fluorescence of this compound occurs at much longer wavelengths than that of anthracene.[8]

Substituents which neither donate nor accept electrons from the aromatic ring (e.g., $-CH_3$, $-NH_3{}^+$) affect the positions of fluorescence and phosphorescence bands only slightly by their polarizing (inductive) effects on the electronic distributions in the ground and excited states of the aromatic ring. The methyl group has a negatively polarizing influence; however, the effect of this group is not clear-cut, since hyperconjugative interactions may be thought of as weak intramolecular charge transfer. Methyl groups generally produce a small shift to longer wavelengths relative to the fluorescences and phosphorescences of the unsubstituted aromatic rings. Ammonium groups produce a positive polarization and may produce small shifts to longer or shorter wavelengths depending on the individual aromatic ring.

If an electron-withdrawing and an electron-donating group are attached to the same aromatic ring, fluorescence and phosphorescence may be viewed as involving transitions between the lone pair orbital of the donor group and the vacant π orbital of the acceptor group (Fig. 3). In this case the energy of the transition is lower than when either the donor or acceptor group alone is attached to the aromatic ring and fluorescence and phosphorescence (as well as absorption) are observed at longer wavelengths than when either group alone is attached to the ring. This is the well-known auxochromic effect and depending on one's point of view either the donor or the acceptor group may be considered the chromophore or auxochrome. If two donor groups or two acceptor groups are attached to the aromatic ring, the more strongly interacting group usually dominates the position of the luminescence bands.

In hydrocarbons which have nonequivalent positions in the aromatic ring (e.g., naphthalene and anthracene) the same substituent in different positions may exert different shifting effects on the luminescence bands of the parent hydrocarbon. For example, fluorescence in β-methylnaphthalene occurs at slightly longer wavelengths than in α-methylnaphthalene. This is a result of the electrical interactions of the substituent with the dipole moments of the excited states of the

[8] T. C. Werner and D. M. Hercules, *J. Phys. Chem.* **73**, 2005 (1969).

aromatic ring. Substituents whose electronic interactions (polarization or charge transfer) are transmitted through the smallest angles with the dipole moment vector of the excited state responsible for luminescence have the greatest effect on the energy of the excited state and thus on the wavelengths of luminescence. In naphthalene, the excited state responsible for fluorescence has its dipole moment lying along the long axis of the molecule. Therefore β substituents affect electronic transitions from this state more than α substituents. However, in some cases, notably in naphthalene, strongly interacting α-type substituents may cause energy level reversals so that fluorescence may actually occur from the excited state which has its dipole moment lying along the short axis of the naphthalene ring. The details of the influence of substituent orientation on the nature of the emitting state are beyond the scope of this work and the reader is referred to Winefordner et al.,[6] Murrell,[7] and Mataga and Kubota[9] for further information.

The relationship between the intensity of fluorescence, the absorber concentration, and the fundamental properties of the absorbing and emitting species is derived from the Beer-Lambert law.

$$I_t = I_0 10^{-\epsilon C l} \tag{3}$$

where I_t and I_0 are the intensities of light transmitted by and incident upon the sample, respectively, and ϵ, C, and l are the molar absorptivity of the absorbing species at the wavelength of excitation, the absorber concentration and the optical depth of the sample, respectively. The intensity of light absorbed is

$$I_a = I_0 - I_t = I_0(1 - 10^{-\epsilon C l}). \tag{4}$$

But of all molecules absorbing, only the fraction ϕ_f will return to the ground state by fluorescence; so that the intensity of fluorescence I_f is related to I_a by

$$I_f = \phi_f I_a = \phi_f I_0(1 - 10^{-\epsilon C l}) \tag{5}$$

Thus, fluorescence intensity is not linear but rather an exponential function of absorber concentration. However, if the sample is excited in such a way that the absorbance ($A = \epsilon C l$) is very small (i.e., < 0.02), then the term $(1 - 10^{-\epsilon C l})$ which is the generating function for the power series

$$2.3\epsilon C l - \frac{(2.3\epsilon C l)^2}{2!} + \frac{(2.3\epsilon C l)^3}{3!}$$

[9] N. Mataga and T. Kubota, "Molecular Interactions and Electronic Spectra." Dekker, New York, 1970.

reduces to $2.3\epsilon Cl$. This situation is favored by low absorber concentration and thus in the limit of low absorber concentration Eq. (5) reduces to

$$I_f = 2.3 I_0 \phi_f \epsilon Cl \tag{6}$$

In Eq. (6) I_0, C, and l are experimental variables but ϕ_f and ϵ are molecular constants; ϵ of the absorbing (ground state) species and ϕ_f of the emitting (excited state) species. ϵ is derived from the quantum mechanical probability per unit time of molecular electronic absorption and has been well described in Mason,[3] Turro,[5] and Murrell.[7] The fraction of excited molecules which fluoresce, ϕ_f, is also called the quantum yield or quantum efficiency of fluorescence. For isolated molecules ϕ_f is defined by

$$\phi_f = \frac{n_f}{n_f + n_d} = \frac{k_f}{k_f + k_d} \tag{7}$$

where n_f and n_d are the numbers of molecules returning from the lowest excited singlet state to the ground state by fluorescence and by all other unimolecular deactivating processes (internal conversion, intersystem crossing), respectively, and k_f and k_d are the corresponding numbers per unit time, which translate into the molecular probabilities or rate constants for fluorescence and all other deactivating processes, respectively. Processes in the lowest singlet state, such as internal conversion and intersystem crossing, which compete with fluorescence, increase n_d and k_d, thereby reducing ϕ_f and the intensity of fluorescence. Substituting alkyl side chains onto aromatic rings or increasing temperature increases the vibrational freedom of the molecule and thereby the probability of internal conversion. Heteroatoms and high atomic number substituents increase spin–orbital coupling and hence the probability of singlet–triplet intersystem crossing. These structural effects thereby tend to diminish the intensity of fluorescence relative to that of the unsubstituted molecules.

The relationship between phosphorescence intensity and the molecular constants of absorption and phosphorescence is similar to the corresponding relationship for fluorescence

$$I_p = \phi_{ST}\phi_P I_0 (1 - 10^{-\epsilon Cl}) \tag{8}$$

where I_P is the intensity of phosphorescence, ϕ_{ST} is the fraction of excited molecules which undergo intersystem crossing from the lowest excited singlet to the triplet state (the intersystem crossing efficiency or quantum yield), ϕ_P is the fraction of molecules which arrives in the triplet state and is deactivated by phosphorescence (the phosphorescence efficiency or quantum yield), and I_0, ϵ, C, and l have the same meanings as in Eq. (3). In the limit of low absorber concentration, Eq. (8) reduces to

$$I_P = 2.3\phi_{ST}\phi_P I_0 \epsilon Cl \tag{9}$$

The intersystem crossing quantum yield is given by

$$\phi_{ST} = \frac{n_{ST}}{n_f + n_d} = \frac{k_{ST}}{k_f + k_d} \tag{10}$$

where n_{ST} and k_{ST} are, respectively, the number and rate constant of molecules populating the triplet state from the excited singlet n_f, n_d, k_f, and k_d have the same meanings as in Eq. (7). Intersystem crossing is a vibrational spin-forbidden process. Thus, substitution in an aromatic ring by alkyl groups and heteroatoms as well as atoms of high atomic number favor a large value of n_{ST} and k_{ST} and therefore of ϕ_{ST} and I_P. The phosphorescence quantum yield is given by

$$\phi_p = \frac{n_p}{n_p + n_d'} = \frac{k_p}{k_p + k_d'} \tag{11}$$

where n_p and n_d' are the numbers of molecules originally populating the triplet state which return to the ground state by phosphorescence and by all other means, respectively, while k_p and k_d' are the corresponding rate constants. Alkyl substituents and elevated temperatures increase vibrational freedom and therefore vibrational deactivation of the triplet (triplet–singlet intersystem crossing). This favors large values of n_d' and k_d' and therefore small values of ϕ_p and I_p. Heteroatoms and atoms of high atomic number increase, by virtue of spin–orbital coupling, the probabilities of all spin-forbidden processes so that n_p, k_p, n_d', k_{ST}, and k_d' are all expected to rise under the influence of these perturbations. In this case, the presence of a heteroatom usually favors a high intensity of phosphorescence at the expense of the intensity of fluorescence. However, this is not universally true. Very heavy atom substituents such as Hg in chelates tend to increase k_d' faster than k_p, so that phosphorescence as well as fluorescence is quenched (diminished in intensity).

B. Heterocyclic Molecules

The introduction of a nitrogen, oxygen, or sulfur atom into an aromatic ring in place of a carbon atom may be treated, from the spectroscopic point of view, as a special substituent effect. By far the greatest amount of study of luminescence properties of heterocycles has been carried out on nitrogen heterocycles. Moreover, the oxygen and sulfur heterocycles whose fluorescence and phosphorescence properties have been studied are usually more complex molecules than simple heterocycles. The nitrogen heterocycles will be considered first.

1. Nitrogen Heterocycles

Simple nitrogen heteroaromatics (those containing one nitrogen heteroatom) may be classified in two categories. Nitrogen atoms occurring in six-membered aromatic rings (e.g., in pyridine, quinoline, and acridine) are trigonally hybridized, σ bonded to two adjacent atoms, contribute one electron to the ring π system and have a nonbonded electron pair residing in an sp^2 orbital projected away from and in the plane of the aromatic ring. A single nitrogen atom in a five-membered aromatic ring (e.g., pyrrole, indole, and carbazole) is also trigonally hybridized, but is σ bonded to three adjacent atoms, one of which is a hydrogen atom and contributes two electrons (its lone pair) to the aromatic π system. Pyrrolic nitrogen atoms therefore do not have a truly nonbonded electron pair. Moreover, pyridinic nitrogen atoms differ from the carbon atoms in their aromatic rings in that the electronegativity of the nitrogen atom is greater than that of carbon. The pyridinic nitrogen atom thus tends to accumulate excess π electronic charge about itself in ground and excited states by virtue of its inductive effect. On the other hand, while the pyrrolic nitrogen atom has a stronger inductive effect than carbon, the donation of the nitrogen lone pair to the aromatic ring represents a stronger electronic effect which leaves the nitrogen atom electron deficient (because each carbon atom contributes only one electron to the π system). In the lower excited states the electron deficiency of the pyrrolic nitrogen atom is even more exaggerated,[10] suggesting electronic behavior similar to that of an arylamine. The differences between the π-electronic properties of pyridinic and pyrrolic nitrogen atoms and the presence of the nonbonded electron pair on the pyridinic nitrogen atom are responsible for the differences between the spectroscopic as well as the chemical properties of aromatic compounds containing one type of nitrogen atom or the other.

  a. Heterocycles Containing Pyridinic Nitrogen Atoms. The orbital containing the nonbonded electron pair of pyridinic type nitrogen atoms is intermediate in energy between the highest occupied π orbital of the

[10] A. C. Capomacchia and S. G. Schulman, *Anal. Chim. Acta* **59**, 471 (1972).

aromatic ring and the pure $2p$ orbital energy. Thus the energy gap between the nonbonded orbital and the lowest unoccupied π orbital is smaller than the gap between the highest occupied and lowest unoccupied π orbitals. As a result, the excited state formed by promotion of a nonbonded electron to the lowest unoccupied π orbital (the lowest $^1n,\pi^*$ state) is lower in energy than the excited state formed by promotion of an electron from the highest occupied to the lowest unoccupied π orbital (the lowest $^1\pi,\pi^*$ state) and is therefore the lowest excited singlet state of the aromatic system. Fluorescence, if it is to occur, must occur from the lowest $^1n,\pi^*$ state. However, because of poor overlap between the nonbonded orbital, which lies in the plane of the ring, and the unoccupied π orbitals, which are directed perpendicular to the plane of the ring, the transitions between the ground singlet state and the lowest $^1n,\pi^*$ state are of low probability. The reciprocal of the probability of transition is the mean lifetime of the excited state, so that a low transition probability translates into a long mean lifetime of the lowest $^1n,\pi^*$ state. As a result of the long lifetime of the $^1n,\pi^*$ state, nonradiative processes such as singlet–triplet intersystem crossing compete more effectively with fluorescence than in the shorter lived $^1\pi,\pi^*$ states. Moreover, the nonbonded orbital is sp^2 hybridized and electrons occupying these orbitals have greater s character than the $p\pi$ electrons. This, along with the greater electronegativity of nitrogen, means that the nonbonded electrons of pyridinic nitrogen atoms spend more time near the nitrogen nucleus than $p\pi$ electrons in homocyclic rings so that pyridinic heterocycles exhibit greater spin–orbital coupling and therefore greater population of the triplet state by intersystem crossing. As a result, the $^1n,\pi^*$ state is efficiently deactivated by intersystem crossing so that molecules having pyridinic nitrogen atoms do not fluoresce or fluoresce very weakly. Pyridine does not fluoresce at all in nonpolar solvents while quinoline fluoresces very weakly in about the same spectral region where naphthalene fluoresces.[11] The fluorescence of acridine is also extremely weak in nonpolar solvents[12] and is similar in position to that of anthracene.

The efficient population of the triplet states of pyridinic heterocycles by intersystem crossing suggests that these molecules should exhibit phosphorescence, and usually they do. However, whereas phosphorescence always originates from the lowest triplet state, it does not always originate from the $^3n,\pi^*$ state even though the lowest excited singlet state is the $^1n,\pi^*$ state. Because the π and π^* orbitals are both projected

[11] B. L. VanDuuren, *Chem. Rev.* **63**, 325 (1963).

[12] B. L. VanDuuren, *Anal. Chem.* **32**, 1436 (1960).

into the same region of space (perpendicular to the plane of the ring) the difference in energy (splitting) between the $^1\pi,\pi^*$ state and the $^3\pi,\pi^*$ state due to differences in electronic repulsions is greater than between the $^1n,\pi^*$ state and the $^3n,\pi^*$ state in which repulsions are smaller since the n and π^* orbitals are projected in different regions of space. As a result, even though the $^1n,\pi^*$ state may lie below the lowest $^1\pi,\pi^*$ state, the large splitting energy between the $^1\pi,\pi^*$ and $^3\pi,\pi^*$ states may result in the $^3\pi,\pi^*$ state lying lower than the $^3n,\pi^*$ state. If this occurs, intersystem crossing from the $^1n,\pi$ state terminating in the $^3n,\pi^*$ state, is followed by internal conversion to the $^3\pi,\pi^*$ state, and then perhaps by phosphorescence. Otherwise phosphorescence occurs directly from the $^3n,\pi^*$ state, and the closer the lowest $^1\pi,\pi^*$ state lies to the $^1n,\pi^*$ state as in the larger aromatic systems, the more likely it is that the $^3\pi,\pi^*$ state will lie lowest. If phosphorescence occurs from the $^3\pi,\pi^*$ state (or fluorescence from the $^1\pi,\pi^*$ state) of a nitrogen hetero-cycle, the emission maximum is likely to lie close to the corresponding emission maximum of the parent homocyclic molecule because the inductive effect of the pyridinic nitrogen heteroatom influences the positions of the π,π^* states only by a small amount. However, if phosphorescence occurs from the $^3n,\pi^*$ state (or fluorescence from the $^1n,\pi^*$ state) the emission maximum of the heterocycle will usually occur at much longer wavelengths than that of the parent hydrocarbon. Thus quinoline[13] and quinoxaline[14] (1,4-diazanaphthalene) phosphoresce from the $^3\pi,\pi^*$ state at low temperatures, with $\lambda_p = 457$ nm and $\lambda_p = 465$ nm, respectively, close to the phosphorescence maximum of naphthalene. Acridine phosphoresces in the same general spectral region as does anthracene. Moreover, phenazine ($\lambda_p = 652$), which is 9,10-diazaanthracene, also shows an anthracene-like phosphorescence.[15] o-Phenanthroline, a diazaphenanthrene, demonstrates a phenanthrene-like fluorescence at $\lambda_f = 358$ nm and phosphorescence at $\lambda_p = 450$ nm.[16,17] Pyridine, however, exhibits neither fluorescence nor phosphorescence. This represents one of the more significant unsolved problems of molecular electronic spectroscopy. It is possible that, in pyridine, the lowest triplet state is efficiently deactivated by radiationless triplet–singlet intersystem crossing to the ground state. The diazabenzenes pyridazine

[13] V. Ermolaev and I. Kotlyar, *Opt. Spectrosc.* **9**, 183 (1960).
[14] J. Vincent and A. Maki, *J. Chem. Phys.* **39**, 3088 (1963).
[15] A. Grabowska and B. Pakula, *Photochem. Photobiol.* **9**, 339 (1969).
[16] S. G. Schulman, P. T. Tidwell, J. J. Cetorelli, and J. D. Winefordner, *J. Amer. Chem. Soc.* **93**, 3179 (1971).
[17] J. Brinen, D. Rosebrook, and R. Hirt, *J. Phys. Chem.* **67**, 2651 (1963).

STEPHEN G. SCHULMAN

(1,2-diaza), pyrimidine (1,3-diaza), and pyrazine (1,4-diaza) fluoresce weakly at 376 nm 326 nm and 327 nm, respectively.[18] Moreover, pyridazine does not phosphoresce at all whereas pyrimidine and pyrazine phosphoresce at 353 nm and 377 nm, respectively, much more intensely than they fluoresce. That the emission maxima are at so much longer wavelengths than those of benzene is indicative of fluorescence from the $^3n,\pi^*$ states of the heterocycles. If the emissions occurred from the π,π^* states of the heterocycles they would lie very close to those of benzene. Sym-triazine phosphoresces (λ_p = 379 nm) from its lowest $^3n,\pi^*$ state and shows no fluorescence.[19] However sym-tetrazine does not phosphoresce and its fluorescence at 580 nm has been shown to originate from the $^1n,\pi^*$ state.[20] It is rather interesting that although 1,10-diazaphenanthrene phosphoresces (λ_p = 450 nm) and fluoresces (λ_f = 358 nm) from π,π^* excited states, 9,10-diazaphenanthrene phosphoresces (λ_p = 580 nm) from a $^3\pi,\pi^*$ state but fluoresces (λ_f = 433 nm) from a $^1n,\pi^*$ excited state.[21]

b. *Heterocycles Containing Pyrrolic Nitrogen Atoms.* Because the lone electron pairs of pyrrolic nitrogen atoms are more like the lone pairs of exocyclic amino groups than like the nonbonded pairs of pyridinic nitrogen atoms, the enhancement of phosphorescence at the expense of fluorescence is not as pronounced in pyrrolic heterocycles as it is in pyridinic heterocycles. The lowest excited states of the simple pyrrolic heterocycles (e.g., pyrrole, indole, and carbazole) are of the π,π^* or intramolecular charge transfer type rather than of the n,π^* type. Pyrrole itself neither fluoresces nor phosphoresces. However, indole and carbazole both fluoresce and phosphoresce intensely. The replacement of the methylene carbon atoms of indene (λ_f = 290 nm) and fluorene (λ_f = 303 nm) to form indole and carbazole, shifts the fluorescences of these compounds to longer wavelengths (for indole in cyclohexane λ_f = 297 nm),[22] for carbazole in cyclohexane λ_f = 332 nm.[22] This is similar to the fluorescence shifts observed when methyl groups are replaced by amino groups in substituted aromatics. However, the phosphorescences of indole (λ_p = 404 nm)[23] and carbazole (λ_p = 407 nm)[24] are at slightly shorter wavelengths than the phosphorescences

[18] B. Cohen and L. Goodman, *J. Chem. Phys.* **46**, 713 (1967).

[19] L. Goodman, *J. Mol. Spectrosc.* **6**, 109 (1961).

[20] M. Chowdhury and L. Goodman, *J. Chem. Phys.* **36**, 548 (1962).

[21] E. Lippert, W. Luder, F. Mall, W. Nagele, H. Boss, H. Prigge, and L. Seipold-Blankenstein, *Angew. Chem.* **73**, 695 (1961).

[22] B. L. VanDuuren, *J. Org. Chem.* **26**, 2954 (1961).

[23] V. Ermolaev, *Opt. Spectrosc.* **11**, 266 (1961).

[24] R. Heckman, *J. Mol. Spectrosc.* **2**, 27 (1958).

of indene (λ_p = 405 nm) and fluorene (λ_p = 421 nm). This may be due to the stronger inductive effect of the nitrogen atom relative to that of carbon and the weak charge tranfer properties of the triplet state.

All nitrogen atoms in pyrrolic (five-membered) rings are not necessarily of the pyrrolic type. For example, the nitrogen atoms in the 3-position of benzimidazole and in the 2-position of indazole are pyridinic in nature, having a nonbonded pair. The inductive effect of the pyridinic nitrogen in the heterocyclic ring of indole is small, usually resulting in a slight shift of the luminescence spectra to shorter wavelengths relative to those of indole (e.g., for benzimidazole λ_f = 290 nm).[25] In this respect it is similar to the effect produced by introducing a second pyridinic nitrogen atom into a six-membered ring. However, because the presence of a pyridinic nitrogen atom in the homocyclic ring of indole produces a polarizing effect which assists the intramolecular transfer of charge between the pyrrolic nitrogen atom and the homocyclic ring, the introduction of a pyridinic nitrogen into any of the various positions of the homocyclic ring of indole generally produces longer fluorescence wavelengths, relative to the fluorescence of indole.

In the diazaindenes (monoazaindoles), $^1n,\pi^*$ states introduced by the presence of the pyridinic nitrogen atoms are higher in energy than the $^1\pi,\pi^*$ (intramolecular charge transfer singlet) states and intense fluorescence is usually observed. However, in the tri- and tetraazaindenes, (e.g., purine) the $^1n,\pi^*$ states lie lowest so that phosphorescence but weak or no fluorescence is most often observed.

2. Oxygen and Sulfur Heterocycles

There is currently very little information concerning the luminescence properties of simple (unsubstituted) oxygen and sulfur heterocycles. Furan, thiophene, pyran, and thiapyran neither fluoresce nor phosphoresce. However, the oxygen and sulfur analogs of carbazole, dibenzofuran, and dibenzothiophene phosphoresce at 408 and 411 nm and the sulfur analog of indole, thianaphthene, at 416 nm, respectively.[24] The wavelength of phosphorescence of dibenzofuran is identical to that of fluorene and similar to that of carbazole. However, the phosphorescences of thianaphthene and dibenzothiophene occur at slightly longer wavelengths, a fact which reflects the effect of the higher energy $3p\pi$ orbital of sulfur (as opposed to the $2\pi p$ orbitals of nitrogen and oxygen) on the energy of the highest occupied molecular orbital.

In contrast to the simple monocyclic heterocycles of oxygen, the mixed nitrogen–oxygen heterocycles are often intensely fluorescent; so

[25] M. Kondo and H. Kuwano, *Bull. Chem. Soc. Japan* **42**, 1433 (1969).

much so that several oxazoles are employed as primary scintillators in radioactivity counting.

3. Substituent Effects in Heterocyclic Molecules

The concepts on which the effects of substituents in aromatic rings on fluorescence and phosphorescence are based were developed earlier in this chapter. Substituents in heterocyclic rings may be either electron donating or electron accepting. The heteroatoms themselves may be thought of as electron withdrawing (as in the case of pyridinic nitrogen atoms) or electron donating (e.g., pyrrolic nitrogen atoms). Thus the introduction of an electron-donating group into an aromatic system containing a pyridinic nitrogen atom or an electron-withdrawing group into an aromatic system containing a pyrrolic nitrogen atom, oxygen atom, or sulfur atom amounts to the addition of an auxochrome to the heterocyclic system. In these types of molecules, the lowest excited singlet state is of the intromolecular charge transfer type (having the characteristics of a $^1\pi,\pi^*$ state) and fluorescence is favored over phosphorescence. For example, in hexane, 8-hydroxyquinoline and 5-hydroxyquinoline both fluoresce at ~ 400 nm but neither of these compounds phosphoresce.[26] In Table I the fluorescence maxima of the

TABLE I

FLUORESCENCE MAXIMA OF THE AMINOQUINOLINES IN
n-HEPTANE AT ROOM TEMPERATURE (298°K)

	λ_f (nm)
2-Aminoquinoline[a]	372
3-Aminoquinoline[b]	372
4-Aminoquinoline[a]	361
5-Aminoquinoline[c]	441
6-Aminoquinoline[d]	381
7-Aminoquinoline[d]	404
8-Aminoquinoline[c]	439

[a] P. J. Kovi, A. C. Capomacchia, and S. G. Schulman, Anal. Chem. **44**, 16 (1972).

[b] S. G. Schulman and A. C. Capomacchia, Anal. Chim. Acta **58**, 91 (1972).

[c] S. G. Schulman and L. B. Sanders, Anal. Chim. Acta **56**, 91 (1971).

[d] S. G. Schulman, K. Abate, P. J. Kovi, A. C. Capomacchia, and D. Jackman, Anal. Chim. Acta **65**, 59 (1973).

[26] M. Goldman and E. L. Wehry, Anal. Chem. **42**, 1178 (1970).

isomeric aminoquinolines are shown. In each case the fluorescence is at longer wavelengths and more intense than the fluorescence of quinoline (or the corresponding naphthylamine). Moreover, the isomers substituted in the homocyclic ring show fluorescence at longer wavelengths than those substituted in the heterocyclic ring of quinoline. This is a result of the transfer of charge to the heterocyclic ring from the homocyclic ring in the lowest excited singlet state of quinoline. Electron-donating substituents in the heterocyclic ring tend to destabilize the excited state dipole while those in the homocyclic ring stabilize it with respect to the ground state. The substantial differences between the fluorescence maxima of the various isomers suggest that fluorescence spectroscopy can be extremely useful in the *in situ* identification of products formed in reactions where several isomers are produced.

Aza substitution in the homocyclic ring of indole amounts to the introduction of an electron-withdrawing substituent auxochromic to the pyrrolic nitrogen atom. Consequently the fluorescences of 4, 5, 6, 7, and 8 azaindole are at longer wavelengths than that of unsubstituted indole.[27]

The introduction of an electron-donating substituent into a heterocyclic ring containing an electron-donating heteroatom affects the fluorescence maximum to only a small extent. The energy of the intramolecular charge transfer transition which determines the position of the fluorescence maximum depends on the energy separation between the lowest unoccupied π orbital and the highest occupied lone pair orbital. As a result, the lone pair orbitals of the donor heteroatom and the substituent group are in competition with each other for the position of the highest occupied orbital. The differences in energy between lone pairs arising from nitrogen, oxygen, and even sulfur (those atoms which are most likely to furnish lone pairs) are small by comparison with the separation between the lone pair orbitals and the lowest unoccupied π orbital. In the hydroxyindoles, for example, the fluorescence maxima occur at wavelengths comparable to or slightly greater than the fluorescence maximum of unsubstituted indole.[28]

If an electron-withdrawing substituent is introduced into an aromatic system containing an electron-withdrawing heteroatom (i.e., a pyridinic nitrogen atom), the fluorescence maximum is expected to occur at longer wavelengths than in the unsubstituted heterocycle. This is a result of the introduction of a vacant π orbital of the substituent, at lower energy than the lowest unoccupied π orbital of the unsubstituted

[27] T. Adler, *Anal. Chem.* **34**, 685 (1962).

[28] S. Udenfriend, D. F. Bogdanski, and H. Weissbach, *Science* **122**, 972 (1955).

heterocyclic ring. Thus while the fluorescence of quinoline lies at about 330 nm, the fluorescence maxima of 2-carbomethoxyquinoline and 8-carbomethyoxyquinoline lie at 395 and 461 nm, respectively.[29] However, in unsubstituted pyridinic heterocycles, the highest occupied orbital is usually a nonbonding orbital, with the result that the lowest excited singlet state is of the $^1n,\pi^*$ type. The introduction of an electron-donating substituent into a pyridinic heterocycle generally results in the lowest excited state being of the intramolecular charge transfer or $^1\pi,\pi^*$ type because the lone pair of the substitutent is generally at somewhat higher energy than the nonbonded pair of the pyridinic nitrogen atom. However, if the substituent is of an electron-withdrawing nature, the vacant π orbital it introduces to the aromatic system does not change the position of the nonbonding orbital of the pyridinic nitrogen atom relative to the energies of the occupied π orbitals of the aromatic system. Moreover, many electron-withdrawing groups such as acetyl, aldehyde, and carboxylate introduce nonbonding orbitals into the aromatic system.

Consequently, in heterocycles containing pyridinic nitrogen atoms and substituted only with electron-withdrawing groups, the lowest excited state is usually of the $^1n,\pi^*$ type. This suggests that fluorescence should be either weak or nonexistent in these compounds, and in fact, the fluorescences of the 2- and 8-carbomethoxyquinolines are very weak. These compounds do, however, exhibit intense phosphorescence, as would be anticipated for compounds having an $^1n,\pi^*$ state as the lowest excited singlet state. Several derivatives of pyridine which contain electron-withdrawing groups have been found to exhibit phosphorescence. For example, pyridine 2-carboxylic acid,[30] pyridine-3-carboxylic acid,[31] pyridine 3-sulfonic acid[32] and 3-acetylpyridine[32] phosphoresce at 410, 400, 411, and 424 nm, respectively. However, none of these compounds demonstrates measurable fluorescence.

While the principles governing the relationships between substituents and the positions of fluorescence spectral bands have been studied in considerable detail, there exists at present a paucity of information concerning the effects of chemical structure, especially positional isomerism, on the positions of phosphorescence spectra. This should prove to be a fertile area for future investigation.

Finally, a word of caution regarding structural interpretations

[29] P. J. Kovi, C. L. Miller, and S. G. Schulman, *Anal. Chim. Acta* **62**, 59 (1972).
[30] S. G. Schulman, unpublished data.
[31] L. V. S. Hood and J. D. Winefordner, *Anal. Biochem.* **27**, 523 (1969).
[32] L. B. Sanders, J. J. Cetorelli, and J. D. Winefordner, *Talanta* **16**, 407 (1969).

based on spectral comparisons is in order. Comparison of luminescence band maxima of different molecules should ideally represent comparison of the differences in electronic energy between the excited states and ground states of the molecules under consideration. However, owing to the vibrational substructure of molecular luminescence bands and the differences in the relative intensities of the vibrational subbands from molecule to molecule, the emission band maxima of different molecules may differ not only in electronic energy but in vibrational energy as well. This situation complicates the interpretation of luminescence spectral differences between molecules in terms of simple differences in molecular electronic structure. It is for this reason that the position of the O–O vibrational subband (the shortest wavelength vibrational feature of the luminescence band) is usually taken to represent the energy of the spectral transition, whenever possible, regardless of whether or not it is the luminescence band maximum. The O–O band of a luminescence band represents a pure electronic transition in the sense that no vibrational energy is included in that transition. It is not, however, always possible to locate the O–O band of a molecular emission band. In particular, many substituted aromatic molecules do not exhibit a distinct vibrational structure but rather have a Gaussian appearance. In these cases comparisons must be made between band maxima in the hope that the maxima under comparison either are the O–O bands or are equally displaced in vibrational energy from their respective O–O bands so that the difference in vibrational energy cancels out in comparison. The margin for error in spectral interpretations arising from comparisons between luminescence bands in which the position of the O–O band is unknown for one or more molecular species is of course greater than when the positions of the O–O bands are known for all species under consideration.

IV. SOLVENT INFLUENCES ON FLUORESCENCE AND PHOSPHORESCENCE SPECTRA

The properties of the solvents in which fluorescence and phosphorescence spectra are observed are of prime importance in determining the spectral positions and intensities with which fluorescence and phosphorescence bands occur. In some cases, especially where heterocyclic molecules are concerned, the solvent may determine whether or not fluorescence or phosphorescence is to be observed at all.

The influences of the solvent on the luminescence spectra are derived from the ways in which the solvent affects the electronic transitions which give rise to the luminescence spectra. These, in turn, are

171

determined by the nature and degree of the interactions of the solvent molecules with the electronic configurations representing the ground and electronically excited states of the luminescing solute molecules. Because of the dramatic effects strongly interacting solvents have on the nature of the states from which fluorescence and phosphorescence occur it will be necessary to consider briefly the effects of the solvent on the absorption spectra and some of the higher excited states as well as the lowest excited states as they exist in hydrocarbon solvents.

The electronic transitions of organic molecules occurring in the near ultraviolet and visible regions of the electromagnetic spectrum fall into three broad categories based on the electronic configurations of the ground and electronically excited states.

(a) $\pi \rightarrow \pi^*$ Transitions involve conjugated molecules which have a ground state in which all π electrons are in the lowest energy π orbitals and an excited state in which a single π electron has been promoted to a higher (antibonding) π orbital. All aromatic molecules exhibit these transitions.

(b) $n \rightarrow \pi^*$ Transitions involve conjugated molecules which have functional groups with lone electron pairs which are essentially localized on single atoms in the ground state. In the excited state one electron from the lone pair has been promoted to an originally unoccupied π orbital of the molecule and is delocalized to some extent. Heterocycles containing pyridinic nitrogen atoms or substituted with certain functional groups (e.g., aldehydes and ketones) exhibit $n \rightarrow \pi^*$ transitions.

(c) Intramolecular charge transfer transitions involve conjugated molecules having functional groups with lone pairs or low-lying localized vacant π orbitals on single atoms (or on the functional group) in the ground state. In the excited state (essentially a π,π^* excited state) however, the localized donor or acceptor orbitals achieve the proper symmetry for interaction with the π electron system of the carbocyclic molecule so that electronic charge from the functional group is substantially delocalized throughout the aromatic system or π electrons from the aromatic system are delocalized and transferred onto the functional group. Heterocycles containing pyrrolic nitrogen oxygen, or sulfur atoms, or substituted with electron donating substituents exhibit these transitions.

Solvent interactions with solute molecules are predominately electrostatic[33] and may be classified as being induced dipole–induced dipole, dipole–induced dipole, dipole–dipole, or hydrogen bonding in nature. In

[33] N. S. Bayliss and E. G. McRae, *J. Phys. Chem.* **58**, 1002 (1954).

addition, the fluidity or rigidity of the solvent and the sizes of the atoms in the solvent molecules also affect the nature and degree of solvent–solute interaction in the various electronic states of the solute.

Induced dipole–induced dipole and dipole–induced dipole interactions are predominant in solutions of nonpolar molecules in polar and nonpolar solvents and of polar molecules in nonpolar solvents. These interactions are extremely weak and account for the small shifts to longer wavelengths of absorption and luminescence spectra of molecules upon going from the gas phase to solutions in nonpolar media. The selection of solutions in nonpolar media as the frame of reference in which to consider the luminescences of isolated molecules, in the previous section, was based upon the negligibility of spectral effects due to induced dipole–induced dipole and dipole–induced dipole interactions in these media. In this section, the spectra in nonpolar, nonhydrogen-bonding solvents will also be taken as the unperturbed spectra of the isolated molecules to which solvent-perturbed spectra will be compared (i.e., hydrocarbon solvents will be assumed not to interact with ground or excited states of the luminescing molecules).

The electrostatic interaction of dipolar solute molecules with dipolar solvent molecules decreases as the third power of the distance between the interacting molecules. As a result, dilution of a solution of a polar solute in a polar solvent with a nonpolar solvent results in an essentially continuous decrease in the dipole–dipole interaction with increasing mole fraction of nonpolar diluent.

Hydrogen-bonding solvents having positively polarized hydrogen atoms are said to be hydrogen bond donor solvents. These interact with the nonbonded and lone electron pairs of solute molecules. Hydrogen-bonding solvents having atoms with lone or nonbonded electron pairs are said to be hydrogen bond acceptor solvents. These interact with positively polarized hydrogen atoms on electronegative atoms belonging to the solute molecules (e.g., in —COOH, —NH$_2$, —OH, —SH).

Hydrogen bonding is a stronger interaction than nonspecific dipole–dipole interaction and is manifested only in the primary solvent cage of the solute. Dilution of a solution of a solute in a hydrogen-bonding solvent by a nonpolar, nonhydrogen-bonding solvent will not normally disrupt the hydrogen bonding in the primary solvent cage. Because most hydrogen-bonding solvents are also polar, and vice versa, hydrogen bonding and nonspecific dipolar interaction are usually both present as modes of solvation of functional molecules. Because of the involvement of nonbonded and lone electron pairs in $n \rightarrow \pi^*$ and intramolecular charge transfer transitions, hydrogen-bonding solvents play a major

173

role in the appearances of these spectra. Polar solvents have a major effect on $\pi \to \pi^*$ and intramolecular charge transfer spectra because of the large dipole moment changes accompanying the corresponding transitions.

Consider the absorption of light by a polar molecule capable of hydrogen bonding, in a solvent of high polarity (dielectric strength) and having both hydrogen bond donor and acceptor properties (e.g., water). In the ground state the molecule will have a solvent cage in which the positive ends of the solvent dipoles will be oriented about the negative ends of the solute dipoles and the negative ends of the solvent dipoles will be oriented about the positive ends of the solute dipoles. Positively polarized hydrogen atoms of the solvent may be oriented toward lone pairs on the solute and acidic hydrogen atoms of the solute may be oriented toward lone pairs on the solvent. The solvent cage is in thermal equilibrium with the ground state electronic distribution of the solute. The light absorption process alters the electronic distribution of the solute so that the electronic dipole moment of the excited molecule is different from that of the ground state molecule. However, the absorption process is so rapid that it terminates with the excited molecule still in the ground state equilibrium solvent cage (the so-called Franck-Condon excited state). If the solute molecule becomes more polar in the excited state, there will be greater electrostatic stabilization of the excited state relative to the ground state by interaction with the polar solvent (Fig. 4). The greater the polarity of the solvent the lower will be the energy of the Franck-Condon excited state. This type of behavior is characteristic of most $\pi \to \pi^*$ and intramolecular charge transfer transitions and is observed as a shift to longer wavelengths of the absorption spectrum with increasing solvent polarity. In the event that the electronic dipole moment is lower in the Franck-Condon excited state than in the ground state, increasing solvent polarity will stabilize the ground state to a greater degree than the excited state and the absorption spectrum will shift to shorter wavelengths with increasing solvent polarity (Fig. 4).

Absorptive transitions of the $n \to \pi^*$ type are usually more affected by hydrogen bond donor properties of the solvent than by solvent polarity *per se*. If a lone pair on a solute molecule is bound by a hydrogen atom of the solvent, the hydrogen-bonding interaction stabilizes the ground state as well as the n,π^* state of the solute. However, since the ground state molecule has two electrons in the nonbonding orbital while the excited state has only one, the stabilization of the ground state is greatest (Fig. 4). As a result the energies of $n \to \pi^*$ absorptions increase

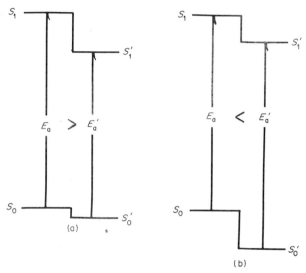

Fig. 4. Illustrating the effect of solvent polarity on the electrostatic energies of the ground and lowest excited singlet states (S_0 and S_1 in hydrocarbon, and S_0 and S_1 in polar solvent) and on the electronic transition energies (E_a and E_a' in hydrocarbon and polar solvent, respectively). (a) The molecule is more polar in S_1 than in S_0; (b) the molecule is less polar in S_1 than in S_0. Molecules having basic electron acceptor substituents in hydrogen bond donor solvents and those having acidic electron donor groups in hydrogen bond acceptor solvents are also represented by (a). Molecules having acidic electron acceptor substituents in hydrogen bond acceptor solvents and basic electron donor groups or nonbonded electron pairs in hydrogen bond donor solvents are also represented by (b).

(the spectra shift to shorter wavelengths) with increasing solvent hydrogen bond donor capacity. In a similar manner, but to a lesser degree, the positive end of the dipole of a polar solvent is capable of producing the same effect upon $n \to \pi^*$ absorption spectra. Hydrogen-bonding solvents also produce a dramatic effect on intramolecular charge transfer absorption spectra. Hydrogen bond donor solvents interacting with lone pairs on functional groups such as carbonyl functions and pyridinic nitrogen atoms which are electron acceptors in the excited state enhance charge transfer by introducing a partial positive charge into the charge transfer acceptor or electronegative group. This interaction stabilizes the charge transfer excited state relative to the ground state so that the absorption spectra shift to longer wavelengths with increasing hydrogen bond donor capacity of the solvent. Following similar lines of reasoning, increasing hydrogen bond donor capacity of the solvent produces shifts to shorter wavelengths when interacting with lone pairs on functional

175

groups which are charge transfer donors in the excited state (e.g., —OH, —NH$_2$). Hydrogen bond acceptor solvents produce shifts to longer wavelengths when solvating hydrogen atoms on functional groups which are charge transfer donors in the excited state (e.g., —OH, —NH$_2$). This is effected by the partial withdrawal of the positively charged proton from the functional group thereby facilitating transfer of electronic charge away from the functional group. Finally, solvation of hydrogen atoms on functional groups which are charge transfer acceptors in the excited state (e.g., —C—OH) inhibits charge transfer by leaving a residual negative charge on the functional group. Thus the latter interaction results in shifting of the absorption spectrum to shorter wavelengths.

It must be borne in mind that the foregoing discussion of solvent-dependent wavelength shifts is based on the consideration of each solvent effect separately. In real molecules, especially polyfunctional molecules, a composite of polarity and hydrogen-bonding effects is usually present. Thus some solvent interactions will tend to produce shifts to longer wavelengths and some to shorter wavelengths in the same molecule. The actual shifts of wavelengths observed, on going from hydrocarbon to polar hydrogen-bonding media, are the results of a composite of all solvent interactions of the molecule.

In rigid solutions, produced by freezing fluid solutions, the effects of the polarity or hydrogen-bonding properties of the solvent on the wavelengths of fluorescence are comparable to those on the wavelengths of the corresponding absorptive transitions, absorption and fluorescence both occurring within the ground state equilibrium solvent cage. Although there has been only limited study of the effects of solvent polarity and hydrogen bonding on the positions of phosphorescence bands, present indications are that the solvent has a much smaller effect on phosphorescence than on absorption or fluorescence. This can be rationalized in terms of the much smaller differences in dipole moment and charge transfer character between the ground and lowest triplet state than between the ground and lowest excited singlet state.[33] Thus the influences of the solvent on the triplet state are much weaker than on the lowest excited singlet state, relative to the ground electronic state. In at least one heterocyclic molecule, the singly charged cation derived from 6-aminoquinoline,[34] the fluorescence maximum ($\lambda =$ 485 nm) in rigid ethanol at 77°K is comparable to the phosphorescence maximum ($\lambda_p = 490$ nm) in the same solvent, indicating that the

[34] S. G. Schulman, K. Abate, P. J. Kovi, A. C. Capomacchia, and D. Jackman, *Anal. Chim. Acta* **65,** 59 (1973).

stabilization of the lowest excited singlet state by interaction with the solvent is almost equal to the sum of the solvent stabilization and spin correlation energies of the lowest triplet state. It is likely that in water or in fluid ethanolic solutions the excited singlet lies below the lowest triplet, a situation which may account for the failure of several molecules which fluoresce at long wavelengths in polar, hydrogen-bonding solvents to phosphoresce.

Although the nature of the rigid solvent has little effect on the wavelength of phosphorescence, it may have a profound effect on the relative intensities of fluorescence and phosphorescence. In molecules whose lowest excited singlet states are of the n,π^* type in nonpolar, nonhydrogen-bonding solvents, fluorescence is usually weak or nonexistent while phosphorescence may be quite intense. The interaction of nonbonded electron pairs with hydrogen bond donor solvents raises the energy of the $^1n,\pi^*$ state whereas the interaction of polar solvent molecules with $^1\pi,\pi^*$ states may lower the energies of these states. If the lowest $^1n,\pi^*$ state lies only slightly below the lowest $^1\pi,\pi^*$ state in nonpolar, nonhydrogen-bonding solvents, the interaction with a polar hydrogen bonding solvent may cause a reversal of the order of these states with respect to the ground state. Since the occurrence of the $^1n,\pi^*$ state as the lowest excited singlet state favors intersystem crossing as a mode of deactivation of the lowest excited singlet state and thereby favors phosphorescence at the expense of fluorescence, reversal of the order of the $^1n,\pi^*$ and $^1\pi,\pi^*$ states in polar, hydrogen-bonding solvents should favor fluorescence at the expense of phosphorescence. This reasoning has been employed to explain the fact that quinoline and acridine are barely fluorescent and intensely phosphorescent in hydrocarbon solvents but fluoresce moderately intensely and show a diminished quantum yield of phosphorescence in ethanol and other hydrogen-bonding media.[35-37]

Solvents which contain atoms of high atomic number (e.g., alkyl iodides) also have a substantial effect on the relative intensities of fluorescence and phosphorescence of solute molecules; however, this effect is not directly related to the polarity or hydrogen-bonding properties of the "heavy atom solvent." Atoms of high atomic number in the solvent cage of the solute molecule enhance spin–orbital coupling in the excited states of the solute. This favors the radiationless population of the lowest triplet state at the expense of the lowest excited

[35] N. Mataga and S. Tsuno, Bull. Chem. Soc. Japan **30**, 368 (1957).
[36] N. Mataga and S. Tsuno, Bull. Chem. Soc. Japan **30**, 711 (1957).
[37] M. A. El-Sayed, J. Chem. Phys. **38**, 2834 (1963).

singlet state. Thus in "heavy atom solvents," all other things being equal, phosphorescence is often more intense and fluorescence always less intense than in solvents of low molecular weight.[38,39]

In fluid solutions, phosphorescence is observed in only a few instances. However, the observation of fluorescence in fluid solutions of aromatic molecules is quite common and the influences of the solvent are far more dramatic than in rigid media.

Subsequent to excitation to the Franck-Condon lowest excited singlet state, the ground state solvent cage reorients itself to conform to the new electronic distribution of the excited molecule. This solvent relaxation process involves reorientation of solute dipoles about new centers of positive and negative charge in the excited molecule and possibly the breaking and making of hydrogen bonds. Because nuclear motions are involved, solvent relaxation is approximately contemporaneous with vibrational relaxation, taking about 10^{-14}–10^{-12} sec, and is rapid by comparison with the lifetime of the lowest excited singlet state ($\sim 10^{-8}$ sec). Consequently, fluorescence originates from the excited solute molecule in a thermally equilibrated solvent cage configuration which is lower in energy than the Franck-Condon excited state and generally even somewhat lower than the vibrationally relaxed, unsolvated or weakly solvated excited molecule. When fluorescence occurs, it terminates in the ground electronic state of the solute molecule, but because of the rapidity of the electronic transition, the molecule is still in the excited state equilibrium solvent cage (higher in energy than the thermally relaxed ground state). Rapid solvent relaxation then occurs (10^{-14}–10^{-12} sec) and the solute molecule ultimately returns to the ground state equilibrium solvent cage. Because the solvent-equilibrated excited state is lower in energy than the Franck-Condon excited state and the Franck-Condon ground state is higher in energy than the solvent-equilibrated ground state (Fig. 5), fluorescence often occurs at considerably longer wavelengths in polar solvents, than would be anticipated purely on the basis of vibrational relaxation. It is for this reason that the O–O bands of fluorescence and absorption often do not coincide.

That the hydrogen-bonding properties of molecules in their lowest excited singlet states are often very different from those in their ground states is reflected in wavelength shifts and in fluorescence intensities on going from nonhydrogen-bonding to hydrogen-bonding solvents.

[38] S. P. McGlynn, M. J. Reynolds, G. W. Daigre, and N. D. Christodouleas, J. Phys. Chem. 66, 2499 (1962).

[39] L. V. S. Hood and J. D. Winefordner, Anal. Chem. 38, 1922 (1966).

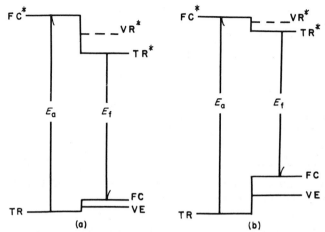

(a)   (b)

FIG. 5. Illustrating the relative magnitudes of solvent relaxation in ground and electronically excited states in (a) a molecule which has a higher dipole moment in the excited state than in the ground state and (b) a molecule which has a higher dipole moment in the ground state than in the excited state. E_a and E_f are the spectral absorption and fluorescence energies. TR and TR* are the energies of the solvent equilibrated ground and excited states, FC and FC* are the energies of the Franck-Condon ground and excited states, respectively. VR* and VE are the energies the thermally relaxed excited state and Franck-Condon ground state would have if vibrational relaxation was the only thermal relaxation mechanism (i.e., if the fluorescence was measured in a rigid matrix so that solvent relaxation was impossible in ground or excited state).

Indole,[40,41] 3-aminoquinoline,[42] 6-aminoquinoline,[34] and 7-aminoquinoline[34] all undergo dramatic shifts of their fluorescence maxima and only small shifts of their absorption maxima on the addition of small amounts of hydrogen-bonding solvents to solutions of these substances in hydrocarbon media. It has been postulated that each of these compounds forms a strong stoichiometric complex with the solvent in the lowest excited singlet state. In Table II, the long wavelength absorption and fluorescence maxima in fluid hydrocarbon, ethanol, and water of the isomeric aminoquinolines which fluoresce in all these media are presented. The shifts of absorption and fluorescence maxima to longer wavelengths on going from hexane to ethanol are, of course, due to stabilization of the lowest excited singlet states of these molecules by strong dipole–dipole interactions and probably also by hydrogen bonding of the lone pairs of the nitrogen atoms in the heterocyclic rings.

[40] M. S. Walker, T. W. Bednar, and R. Lumry, J. Chem. Phys. 45, 3455 (1966).
[41] M. S. Walker, T. W. Bednar, R. Lumry, and F. Humphries, Photochem. Photobiol. 14, 147 (1971).
[42] S. G. Schulman and A. C. Capomacchia, Anal. Chim. Acta 58, 91 (1972).

TABLE II

MAXIMA OF THE LONGEST WAVELENGTH ABSORPTION BANDS (λ_a) AND FLUORESCENCE
BANDS (λ_f) OF THE 2-, 3-, 4-, 6-, AND 7-AMINOQUINOLINES IN n-HEXANE, ETHANOL, AND
WATER. MAXIMA ARE GIVEN IN NM.

	n-Hexane		Ethanol		Water	
	λ_a	λ_f	λ_a	λ_f	λ_a	λ_f
2-Aminoquinoline	332	372	332	380	331	396
3-Aminoquinoline	339	373	352	407	308	411
4-Aminoquinoline	298	362	319	357	312	369
6-Aminoquinoline	340	383	354	443	340	459
7-Aminoquinoline	347	397	357	443	345	455

However, on going from ethanol to water it is apparent that the fluorescence maxima shift to longer wavelengths but the corresponding absorption maxima shift to shorter wavelengths even though water is more polar and a stronger hydrogen bond donor than ethanol. The answer to this apparent dilemma lies in the fact that, as well as the nonbonded pair of the heterocyclic nitrogen atom, the lone pair of the exocyclic amino group also is capable of forming a hydrogen bond which will be stronger with water than with ethanol. The hydrogen bond between water and the lone pair of the amino group substantially destabilizes the Franck-Condon excited state with respect to the thermally equilibrated ground state by impeding the loss of charge from the amino group accompanying light absorption. If this destabilization of the Franck-Condon excited state is greater in magnitude than its stabilization by dipole–dipole interaction and hydrogen bonding at the heterocyclic nitrogen, the observed shifts to shorter wavelengths of the absorption spectra on going to the stronger hydrogen bond donor solvent can be rationalized. However, in the lowest excited singlet state, the amino group is not a good hydrogen bond acceptor because it is charge deficient. The excited state equilibrium solvent cage of an aminoquinoline probably does not have hydrogen bonding to electrons on the amino group as a mode of solvation. As a result, the effect which caused the shift of the absorption maxima or the aminoquinolines to shorter wavelengths on going to the stronger hydrogen bond donor solvent is not present in the molecular species from which fluorescence originates. Thus the increase in polarity and hydrogen bond donor strength attained in going from alcohol to water produces a shift of the fluorescences of the aminoquinolines to longer wavelengths.

5-Hydroxyquinoline,[26] 8-hydroxyquinoline,[26] 5-aminoquinoline,[43] and 8-aminoquinoline[2] fluoresce in solvents of poor hydrogen-bonding

[43] S. G. Schulman and L. B. Sanders, *Anal. Chim. Acta* **56**, 91 (1971).

capability, but their fluorescences are quenched with increasing hydrogen bond donor strength of the solvent. In 5- and 8-aminoquinoline, fluorescence is observed in frozen solutions in strongly hydrogen bonding media (e.g., water) but fluorescence is lost as soon as the solutions thaw to fluidity. This observation indicates that the hydrogen bonding present in the ground state equilibrium solvent cage is not responsible for quenching of fluorescence and that the quenching mechanism is established only after solvent relaxation occurs in the excited state. The mechanism of quenching is not certain but has been postualted to involve thermal deactivation of the lowest excited singlet state resulting from coupling of the excited solute to the solvent structure by hydrogen bonding.

In 8-hydroxyquinoline, replacement of the phenolic proton by a methyl group to yield 8-methoxyquinoline results in fluorescence in hydrogen-bonding as well as nonhydrogen-bonding solvents. This has prompted some investigators[26] to interpret the quenching of 8-hydroxyquinoline fluorescence in water in terms of hydrogen bonding between the phenolic proton and the solvent, in the excited state. Other workers[44-46] have interpreted this phenomenon as the result of tautomerism of the neutral molecule in the lowest excited singlet state to form a nonfluorescent zwitterion. In 8-methoxyquinoline, phototautomerism is impossible; thus the fluorescence of the neutral species persists in all solvents. Of course, it is possible that the intermolecular hydrogen bonding and tautomerism processes, in the excited state, are both responsible for the quenching of 8-hydroxyquinoline fluorescence in hydrogen-bonding solvents.

The great sensitivity of fluorescence maxima and fluorescence intensities to solvent properties suggests the utility of fluorescent molecules as probes of solvent structure.

V. THE INFLUENCE OF ACIDITY OR BASICITY ON MOLECULAR LUMINESCENCE

A. General Considerations

Heteroatoms frequently impart basic properties (by virtue of the presence of lone or nonbonded electron pairs) or acidic properties (by virtue of acidic hydrogen atoms attached to electronegative heteroatoms) to heterocyclic molecules. Moreover, the basic or acidic properties of exocyclic functional groups are often affected by the presence

[44] R. E. Ballard and J. W. Edwards, *J. Chem. Soc.* 4868 (1964).
[45] S. G. Schulman and Q. Fernando, *Tetrahedron* **24**, 1777 (1968).
[46] S. G. Schulman, *Anal. Chem.* **43**, 285 (1971).

of heteroatoms in aromatic rings. Protolytic dissociation may be regarded as an extreme form of interaction with a hydrogen bond acceptor solvent, whereas protonation of a heteroatom or functional group may be regarded as an extreme case of interaction with a hydrogen bond donor solvent. Consequently, the spectral shifts observed when basic groups are protonated or when acidic groups are dissociated are greater in magnitude but qualitatively similar to the shifts resulting from interaction of lone or nonbonded electron pairs with hydrogen bond donor solvents or from interaction of acidic hydrogen atoms with hydrogen bond acceptor solvent, respectively. Thus the protonation of electron-withdrawing groups such as carboxyl, carbonyl, and pyridinic nitrogen results in shifts of the absorption, fluorescence, and phosphorescence spectra to longer wavelengths whereas the protonation of electron-donating groups such as the amino group produces spectral shifts to shorter wavelengths. The protolytic dissociation of electron-donating groups such as hydroxyl, sulfhydryl, or pyrrolic nitrogen produces spectral shifts to longer wavelengths and the dissociation of electron-withdrawing groups, such as carboxyl, produces shifting of the electronic spectra to shorter wavelengths.

The smaller dipole moment in the triplet state compared with the lowest excited singlet state results in smaller environmental perturbations on phosphorescence than on fluorescence or absorption spectra. As in the case of solvent effects, the influence of the state of ionization of a luminescing molecule is much smaller on the position of the phosphorescence band than on the fluorescence band.

Anomalies in the pH dependences of absorption, fluorescence, or phosphorescence spectra of classes of compounds whose spectral behavior is well established may indicate peculiarities in molecular electronic structure. For example, on the basis of the successive shifts of the absorption and fluorescence spectra of the 3-, 5-, 6-, 7-, and 8-hydroxyquinolines[44-48] and the anomalous, by comparison, sequence of spectral shifts in 2- and 4-hydroxyquinoline,[49] it was possible to support evidence from infrared spectroscopy[50,51] for the quinolone (lactam) structures of the uncharged 2- and 4-isomers. The uncharged forms of the 3-, 5-, 6-, 7-, and 8-isomers, by contrast, exist either as

[47] J. C. Haylock, S. F. Mason, and F. E. Smith, *J. Chem. Soc.* 4897 (1963).

[48] S. F. Mason, J. Philp and B. E. Smith, *J. Chem. Soc. A* 3051 (1968).

[49] S. G. Schulman, A. C. Capomacchia and B. Tussey, *Photochem. Photobiol.* 14, 733 (1971).

[50] G. F. Tucker and J. L. Irvin, *J. Amer. Chem. Soc.* 73, 1923 (1951).

[51] S. F. Mason, *J. Chem. Soc.* 4874 (1957).

neutral phenols or as zwitterionic species. These are no subtle differences as the quinolone species behave chemically as carboxamides whereas the neutral and zwitterionic species are phenolic in chemistry. Similarly, anomalous acidity-dependent behavior of the absorption and fluorescence spectra of 2- and 4-aminoquinoline[52] has shown that the singly charged cations derived from the 2- and 4-isomers have the molecular electronic structures of singly protonated cyclic amidines whereas the singly protonated 3-, 5-, 6-, 7-, and 8-aminoquinolines[34,42,43] behave as typical heterocyclic arylamines. An anomalous sequence of pH-dependent shifts of the fluorescence spectra of the antimalarial pamaquine, which is derived from 8-aminoquinoline, was employed to show the presence of an intramolecular hydrogen bond between the nitrogen atom of the quinoline ring and the ammonium group of the alkyl side chain substituted onto the 8-amino group.[53] These are but a small sampling of the potential applications of luminescence spectroscopy to the solution of problems of molecular electronic structure.

One of the more exciting aspects of the study of environmental effects on luminescence spectra is related to the use of the luminescence spectra as indicative of the acid–base photochemistry of the lowest excited singlet and triplet states. This is especially interesting where fluorescence in fluid solutions is concerned, because the rates of many protolytic reactions are comparable to rates of fluorescence, resulting in the obtainability of kinetic information from fluorescence intensity–pH profiles. The acid–base chemistry of electronically excited molecules and its relationship to fluorescence and phosphorescence spectroscopy will now be examined in some detail.

B. Fluorescence and the Acid–Base Chemistry of the Lowest Excited Singlet State

The excitation of a molecule from its ground electronic state to a higher electronic state is accompanied by a change in the electronic dipole moment of the molecule. Accordingly, the distribution of electronic charge in the excited molecule is different from that in the ground state and the excited molecule will have different chemical properties from those of the ground state molecule. In almost every case of fluorescence known, emission originates from the lowest excited singlet state. Thus it is the electronic distribution or the photochemistry of the lowest excited singlet state which is of concern here.

Among the chemical properties of the lowest excited singlet state

[52] P. J. Kovi, A. C. Capomacchia, and S. G. Schulman, *Anal. Chem.* **44**, 16 (1972).
[53] S. G. Schulman and K. Abate, *J. Pharm. Sci.* **61**, 1576 (1972).

which are different from those of the ground state are acidity and basicity of the excited molecule. If the electron density in the region of an acidic or basic group is lower in the excited state than in the ground state, the molecule will be a stronger acid or a weaker base in the excited state. Conversely, if charge transfer occurs to an acidic or basic group, on excitation, the excited molecule will be a weaker acid or a stronger base than in the ground state.

The lifetimes of molecules in the lowest excited singlet state (the average time a molecule spends in the excited state before fluorescing) are typically of the order of 10^{-10}–10^{-8} sec. Typical rates of protolytic reactions are from 10^{11} to 10^8 M^{-1} sec^{-1}. Consequently, excited state prototropic reactions may be much slower, much faster, or competitive with radiative deactivation of the excited molecules. Whether or not protonation or dissociation can occur during the lifetime of the excited state depends on which of the latter three circumstances is extant.

1. Excited State Prototropism Is Much Slower Than Fluorescence

In this case, if the acid form of the analyte is excited, the excited acid will fluoresce before it can convert to the excited conjugate base. The same argument applies to the excitation of the conjugate base species. Thus, which species emits is governed only by which species absorbs, and the relative intensities of emission from acid and conjugate base are determined exclusively by the molar absorptivities of acid and conjugate base at the wavelength of excitation and by the thermodynamics (pK_a) of the ground state prototropic reaction. For example, the ground state pK_a of 3-methoxypyridine, as determined by potentiometry or absorptiometry, is 4.9. The quenching of the ultraviolet fluorescence of the 3-methoxypyridinium ion with increasing pH also yields a pK_a value of 4.9.[54]

An apparent exception to this is sometimes observed when two functional groups, one an electron acceptor and the other an electron donor, are situated ortho to one another in an aromatic ring, with an intramolecular hydrogen bond existing between the two groups in the ground state. Electronic excitation results in the electron acceptor group becoming more basic and the electron donor group more acidic in the lowest excited singlet state. In this case the intramolecularly hydrogen-bonded proton may be completely transferred from the electron donor to the electron acceptor group within the time of a single molecular vibration after absorption ($\sim 10^{-14}$ sec). For example, the anomalously long wavelength of fluorescence of salicylic acid, relative

[54] J. W. Bridges, D. S. Davies, and R. T. Williams, *Biochem. J.* **98**, 451 (1966).

to that of *o*-methoxybenzoic acid is due to emission from the excited salicylic acid zwitterion.[55] However, this is an intramolecular photo-tautomerization reaction rather than a diffusion-limited protonation or dissociation, and the fluorimetric titration behavior of salicylic acid is still that of the ground state neutral molecule.

2. Excited State Prototropism Is Much Faster Than Fluorescence

If the rates of dissociation of the excited acid and of protonation of the excited conjugate base are much greater than the rate of deactivation of the lowest excited singlet states by fluorescence, prototropic equilibrium in the lowest excited singlet state will be achieved.[56] In this event it is the dissociation constant of the excited state prototropic reaction (pK_a^*) which predominantly determines the fluorescence behavior of the analyte. Since the electronic distribution of an electronically excited molecule is generally different from that of its ground state, the pK_a^* of the excited acid is usually very different from the pK_a of the ground state acid. Differences between pK_a^* and pK_a are commonly six or more orders of magnitude and differences of seventeen or more orders of magnitude are not rare.[57] The difference between pK_a^* and pK_a means that the conversion of acid to conjugate base in the excited state occurs in a pH region different from the corresponding ground state reaction. Thus the absorption spectrum of acridine ($pK_a \sim 5.6$) at pH 5 shows the absorption bands originating from both neutral molecule and the acridinium cation while the fluorescence spectrum consists solely of the emission from the excited acridinium cation. However, at pH 10, the absorption spectrum is that of the neutral molecule alone while the fluorescence spectrum shows the blue emission of the excited neutral molecule ($pK_a^* \sim 10$) and the green fluorescence of the excited cation.[58]

When prototropic equilibrium is attained within the lifetime of the lowest excited state and one or both members of the conjugate pair are fluorescent, the determination of pK_a^* is relatively simple matter and is analogous to the spectrophotometric determination of a pK_a. A series of buffer solutions each containing the same amount of analyte is prepared. The fluorescence spectrum of each sample is recorded and the positions of the emission maxima of acid and conjugate base noted. A plot of fluorescence intensity (or relative quantum yield) of either

[55] P. J. Kovi, C. L. Miller, and S. G. Schulman, *Anal. Chim. Acta* **61**, 7 (1972).
[56] A. Weller, *Progr. Reaction Kinet.* **1**, 187 (1961).
[57] S. G. Schulman and Q. Fernando, *J. Phys. Chem.* **71**, 2668 (1967).
[58] N. Mataga, Y. Kaifu, and M. Koizumi, *Bull. Chem. Soc. Japan* **29**, 373 (1956).

species as a function of pH or Hammett acidity allows a rapid estimate of pK_a^* as the value of the pH or Hammett acidity at the midpoint of the fluorimetric titration curve. However, if both acid and base fluoresce and their fluorescence bands overlap, a correction for the overlap of the fluorescence of the conjugate species must be applied before the fluorimetric titration curve is plotted (see, for example, Schulman et al.[16]) so that the ordinate of the plot corresponds to the fractional concentration of excited acid or base alone. Alternatively, pK_a^* may be obtained from the Henderson-Hasselbach equation in the form

$$pK_a^* = pH - \log \frac{\Phi_A{}^\circ - \Phi_A}{\Phi_A - \Phi_B{}^\circ} \tag{12}$$

where Φ_A is the relative fluorescence quantum yield at any point in the fluorimetric titration curve at the analytical wavelength of fluorescence, and $\Phi_A{}^\circ$ and $\Phi_B{}^\circ$ are, respectively, the limiting relative fluorescence quantum yields when $pH \ll pK_a^*$ and when $pH \gg pK_a^*$ at the analytical wavelength.

3. Excited State Prototropism Is Competitive with Fluorescence

If the rates of proton transfer between the excited molecule and the solvent are comparable to the rates of deactivation of acid and conjugate base by fluorescence, the variations of the quantum yields of fluorescence of acid and conjugate base with pH will be governed predominantly by the kinetics of the excited state prototropic reaction. Weller[56,59] has employed simple steady state kinetics to show that, to a first approximation, the variation of relative fluorescence quantum yield of an excited acid with hydrogen ion concentration is given by

$$\frac{\Phi_A}{\Phi_A{}^\circ} = \frac{1 + \overleftarrow{k}\tau_B[H^+]}{1 + \overrightarrow{k}\tau_A + \overleftarrow{k}\tau_B[H^+]} \tag{13}$$

where Φ_A is the relative fluorescence quantum yield at any point in the fluorimetric titration curve of the acid, $\Phi_A{}^\circ$ is the limiting relative fluorescence quantum yield (i.e., $\Phi_A = \Phi_A{}^\circ$ when $pH \ll pK_a^*$), and τ_A and τ_B are the lifetimes of the lowest excited singlet states of acid and conjugate base, respectively. $[H^+]$ is the hydrogen ion concentration and $\overrightarrow{k}$ and $\overleftarrow{k}$ are the rate constants of dissociation of the acid and protonation of the conjugate base, in the lowest excited singlet state, respectively. Correspondingly, if Φ_B is the relative fluorescence quantum

[59] A. Weller, Z. Elektrochem. 56, 662 (1952).

186

yield at any point in the fluorimetric titration of the conjugate base, and $\Phi_B{}^\circ$ is the limiting relative fluorescence quantum yield of the conjugate base ($\Phi_B = \Phi_B{}^\circ$ when $pH \gg pK_a$), then

$$\frac{\Phi_B}{\Phi_B{}^\circ} = \frac{\overleftarrow{k}\tau_A}{1 + \overrightarrow{k}\tau_A + \overleftarrow{k}\tau_B[H^+]} \qquad (14)$$

However, Eqs. (13) and (14) are derived with the assumption that all of the excited molecules enter into the excited state prototropic reaction (i.e., steady state conditions prevail). If it is taken into account that only a fraction (W) of the excited molecules enter into the excited state prototropic reaction while the remaining fraction ($1 - W$) fluoresces without reacting, the relationships for the variations of relative fluorescence intensity with hydrogen ion concentration become

$$\frac{\Phi_A}{\Phi_A{}^\circ} = (1 - W) + W\frac{1 + \overleftarrow{k}\tau_B[H^+]}{1 + \overrightarrow{k}\tau_A + \overleftarrow{k}\tau_B[H^+]} \qquad (15)$$

for the excited acid, and

$$\frac{\Phi_B}{\Phi_B{}^\circ} = \frac{W\overrightarrow{k}\tau_A}{1 + \overrightarrow{k}\tau_A + \overleftarrow{k}\tau_B[H^+]} \qquad (16)$$

for the excited base. W represents the probability that the excited acid will encounter a proton within its diffusion volume during the lifetime of the excited state and may be evaluated from the thermodynamic properties of the solution.[56]

The effect of the kinetic dependence of excited state prototropism on the fluorimetric titration curve is to stretch the region of appreciable dependence of fluorescence intensity on pH over a wider range than would be observed if the protonation kinetics were very fast compared with fluorescence. The faster the rate of the excited state protonation relative to the rate of fluorescence, the steeper will be the fluorimetric titration curve. In the limit of very fast prototropism, the excited state $pK_a{}^*$ defines the shape of the curve while in the limit of very slow prototropism, relative to fluorescence, the ground state pK_a determines the shape of the fluorimetric titration curve (Fig. 6).

If τ_A and τ_B can be measured, Eqs. (13) and (14) [or (15) and (16)] can be solved analytically or graphically for $\overrightarrow{k}$ and $\overleftarrow{k}$. $pK_a{}^*$ can then be calculated from, $pK_a{}^* = -\log \overrightarrow{k}/\overleftarrow{k}$.

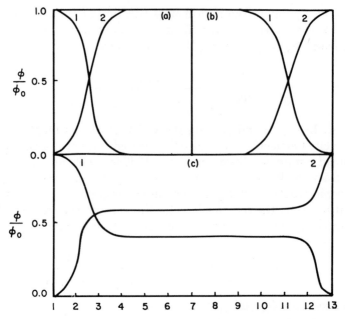

FIG. 6. Variation of the relative quantum yield of fluorescence (ϕ/ϕ_0) of a hypothetical aromatic acid (1) which is more acidic in the lowest excited singlet state ($pK_a = 11.0$, $pK_a^* = 1.5$) and of that of its conjugate base (2) when proton exchange in the lowest excited singlet state is (a) much faster than, (b) much slower than, or (c) comparable in rate to the rate of fluorescence of acid or conjugate base. The flatness of the curves in (c), in the mid-pH region, results from the low rate of protonation of the conjugate base in regions where [H$^+$] is small, making the relative quantum yields essentially independent of pH in these regions. If this were not so, curve (1) in (c) would fall and curve (2) would rise continuously with increasing pH.

4. Calculation of pK_a from Spectral Shifts

As an alternative to the method of fluorimetric titration, pK_a^* values can be evaluated from a thermodynamic cycle (Fig. 7) due to Förster.[60] From Fig. 7 it can be seen that there are two mechanistically different but energetically equivalent ways of converting the ground state acid A to the excited base B*. The first pathway consists of the absorption of electronic energy E_A by A to form A*, followed by dissociation of A* to B* with attendant enthalpy of dissociation ΔH^*. The total energy involved in the latter process is thus $E_A + \Delta H^*$. The second pathway from A to B* entails dissociation of A to B in the ground state, accompanied by the enthalpy of dissociation ΔH, followed by

[60] T. Förster, Z. Electrochem. **54**, 42 (1950).

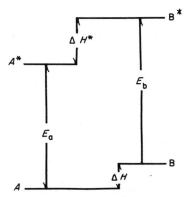

FIG. 7. The Förster cycle: A, A*, B, and B* are the ground and excited states of the conjugate acid and the ground and excited states of the conjugate base, respectively. E_a and E_b are the respective electronic transition energies between A and A* and between B and B*. ΔH and ΔH^* are the respective enthalpies of dissociation in ground and electronically excited states.

excitation of B to B*, the electronic energy E_B being absorbed. The second mechanism requires a total energy of $E_B + \Delta H$. Since both pathways are thermodynamically equivalent,

$$E_A + \Delta H^* = E_B + \Delta H \qquad (17)$$

In Eq. (17) E_A and E_B are, respectively, equivalent to $Nh\nu_A$ and $Nh\nu_B$, where N is Avogadro's number, h is Planck's constant, and ν_A and ν_B are the frequencies of radiation involved in the transitions between A and A* and B and B*, respectively. If it is assumed that the entropies of protonation in ground and excited states are identical, then Eq. (17) becomes

$$\Delta G - \Delta G^* = Nh(\nu_A - \nu_B) \qquad (18)$$

from which it immediately follows that

$$pK - pK^* = \frac{Nh}{2.303RT}(\nu_A - \nu_B) \qquad (19)$$

where R is the universal gas constant and T is the absolute temperature. Finally, if the constants are expressed in cgs units, at 25°C we have

$$pK - pK^* = 2.10 \times 10^{-3}(\bar{\nu}_A - \bar{\nu}_B) \qquad (20)$$

where $\bar{\nu}_A$ and $\bar{\nu}_B$ are the wavenumbers (cm^{-1}) of the transitions from A to A* and B to B*, respectively ($\bar{\nu}_A = 1/\lambda_A$ and $\bar{\nu}_B = 1/\lambda_B$). Consequently, if the transition energies of acid conjugate base are known and the ground state pK_a can be evaluated, the excited state pK_a^* for the corresponding equilibrium can be calculated.

The accuracy with which pK_a^* can be calculated from the Förster cycle is usually established by comparison with pK_a^* values obtained from fluorimetric titration and is subject to the validity of the assumption of equal entropies of protonation in ground and electronically excited states and to the way in which $\bar{\nu}_A$ and $\bar{\nu}_B$ are chosen. These subjects are worthy of some discussion. In aromatic structures in which gross changes in molecular electronic structure do not occur subsequent to excitation, the entropies of protonation in ground and excited states, which are predominantly configurational, are not very different.[61] For example, the difference between the ground and excited state entropies of protolytic dissociation of the β-naphthylammonium ion which arises from rehybridization of the amino group of the conjugate base in the excited state has been shown both experimentally and by statistical mechanical considerations to incur an error of no more than 0.2 log unit in the calculation of pK_a^* from the spectral shift accompanying dissociation.[62]

The quantities $\bar{\nu}_A$ and $\bar{\nu}_B$ in Eq. 20 represent the wavenumbers of "pure" electronic transitions in acid and conjugate base so that $\bar{\nu}_A - \bar{\nu}_B$ should not contain any vibrational energy. To a first approximation, $\bar{\nu}_A$ and $\bar{\nu}_B$ may be taken from either the long wavelength absorption bands of acid and conjugate base, respectively, or if both members of the conjugate pair are fluorescent, from the fluorescence bands of acid and conjugate base, respectively. To avoid inclusion of vibrational energy in the Förster cycle calculation, $\bar{\nu}_A$ and $\bar{\nu}_B$ should be taken at the 0–0 vibronic bands of absorption or fluorescence of acid and conjugate base, respectively. However, in most compounds of interest, the positions of the 0–0 bands are generally not known, especially at room temperature. Levshin[63] has shown that if the spacing between vibrational sublevels in ground and electronically excited states of the same molecule is identical, an approximate mirror image relationship exists between the fluorescence spectrum and the long wavelength absorption band of a fluorescing molecule. If a mirror image relationship exists between absorption and fluorescence, the band maxima of fluorescence and absorption are equally displaced, in energy, from the 0–0 band (the 0–0 band represents the long wavelength limit of absorption and the

[61] H. H. Jaffé and H. L. Jones, *J. Org. Chem.* **30**, 964 (1965).

[62] S. G. Schulman and A. C. Capomacchia, *Spectrochim Acta* **28A**, 1 (1972).

[63] W. L. Levshin, *Z. Phys.* **43**, 230 (1931).

short wavelength limit of fluorescence). Thus by averaging the energies (or wavenumbers) of the fluorescence and long wavelength absorption maxima, the position of the O–O band could be approximated. If both members of a conjugate pair are fluorescent, it is then possible to calculate the wavenumbers of the O–O band for each and apply the Förster cycle. Probably more pK_a^* values in the literature have been obtained by the latter means than by any other.

The calculation of pK_a^* values from the Förster cycle employing the average of absorption and fluorescence maxima is applicable only to those conjugate pairs in which both members are fluorescent. However, if the assumption is made that the vibrational spacings in ground and excited states of both members of a conjugate pair are identical, then the absorption band maxima of acid and conjugate base will be equally displaced from the respective O–O bands.[61] In this case, the absorption maxima may be used in the Förster cycle to calculate pK_a^*; the vibrational discrepancies dropping out in the difference $\bar{\nu}_A - \bar{\nu}_B$. It should be noted that in this case it is not necessary for either member of a conjugate pair to be fluorescent in order to calculate a pK_a^* although such a pK^* is meaningless from the kinetic point of view since lack of fluorescence implies that the excited molecule is deactivated in a time too short for excited state prototropism to occur. However, the thermodynamic significance of such a pK_a^* is presumably the same as if excited state prototropism did occur.

Calculations of excited state pK_a^* values from the Förster cycle employing absorption data alone entail another source of error apart from vibrational considerations. If appreciable solvent relaxation occurs subsequent to excitation and the energies of solvent relaxation are unequal for acid and conjugate base, pK_a^* values determined by fluorimetric titration or from shifts of the fluorescence spectrum alone in conjunction with the Förster cycle will reflect the effect of solvent relaxation since the relaxation process affects the energies of the species involved in the excited state equilibrium. However, because absorption occurs in a time which is short compared with the time required for solvent relaxation the absorption spectra will not reflect the effect of solvent relaxation and pK_a^* values calculated from absorption data alone may be in error. In the latter regard, it has been found[16] that solvent relaxation errors may also be present in pK_a^* values determined from the Förster cycle employing the average of absorption and fluorescence maxima. The pK_a^* value of doubly protonated o-phenanthroline determined by fluorimetric titration indicated that doubly protonated phenanthroline is a stronger acid in the lowest excited singlet state than

191

in the ground state.[16] The value of pK_a^* obtained from the Förster cycle for the latter acid employing fluorescence shifts alone was in qualitative agreement with the value obtained from the fluorimetric titration. However, the pK_a^* obtained from absorption data alone and from the averaging of absorption and fluorescence data indicated that o-phenanthroline dication was a weaker acid in the excited state than in the ground state. In nonpolar solvents, where solvent relaxation is unimportant, fluorescence shifts, absorption shifts, and the averages of absorption and fluorescence shifts all indicated that o-phenanthroline dication was a weaker acid in the excited state. These results serve to illustrate that in some cases the effect of solvent relaxation in the excited state may actually reverse the effect of the electronic shift occurring on excitation in determining the acidity of the excited molecule. Successive pK_a^* values for several diazanaphthalenes and for phenazine have been calculated using the Förster cycle and the averages of absorption and fluorescence maxima.[64,65] In all cases the diazaheterocycles are more basic in the excited state with respect to the first protonation. However, whether the second protonation is easier or more difficult in the excited state appears to depend on the molecular symmetry in some cases.

Correct application of the Förster cycle to the study of excited state prototropism involves several subtleties of which the investigator should be aware. The pK_a and pK_a^* values in Eq. (20) correspond to the same equilibria (i.e., dissociation from the same functional group) in ground and excited states. In polyfunctional molecules, dissociation frequently occurs from different groups in ground and excited states, as a result of the difference in charge distributions in those states. For example, consider a solution of 4-methyl-7-hydroxycoumarin (β-methylumbelliferone). In alkaline solutions this molecule dissociates in the ground state from the hydroxy group. In concentrated acid solutions in the ground state, the carbonyl group becomes protonated. However, in the lowest excited singlet state, uncharged β-methylumbelliferone exists as a zwitterion, with phenolic group dissociated and carboxyl group protonated.[66] In the ground state, the zwitterion is not measurably present. In the excited state the neutral molecular species is not measurably present. The dissociation of the excited zwitterion in alkaline solutions occurs from the protonated carbonyl group while the protonation of the excited zwitterion in acid solutions occurs at the dissociated

[64] A. Grabowska, B. Pakula, and J. Pancir, *Photochem. Photobiol.* **10**, 415 (1969).
[65] A. Grabowska and B. Pakula, *Photochem. Photobiol.* **9**, 339 (1969).
[66] G. J. Yakatan, R. J. Juneau, and S. G. Schulman, *Anal. Chem.* **44**, 1044 (1972).

phenolic group. Clearly, the ground state and excited state pK_a values will not correspond to the same equilibria, nor will absorption and fluorescence spectra correspond to the same uncharged species. Consequently the Förster cycle cannot be meaningfully applied to a comparison of the ground and excited state equilibria of β-methylumbelliferone or other substances that demonstrate phototautomerism during the lifetime of the lowest excited singlet state. The $\bar{\nu}_A$ and $\bar{\nu}_B$ terms in the Förster cycle are presumed to belong to transitions from the ground states of acid and conjugate base, respectively, to corresponding electronically excited states. The term "corresponding" as employed here means that the electronic configurations and angular momentum quantum numbers of the excited states of acid and conjugate base are identical (i.e., the presence or absence of the dissociating proton is regarded as producing a weak perturbation). Occasionally it may be found that the excited states in acid and conjugate base do not correspond. Thus dissociation in the excited state entails an electronic transition as well as a chemical reaction. This phenomenon will usually introduce a large entropy (associated with configurational change) error into the Förster cycle calculation.[61] For example, the molecule 3,4-benzocinnoline[67] fluoresces from an n,π^* lowest excited singlet state. As acid is added to a solution of 3,4-benzocinnoline the fluorescence is quenched. Fluorimetric determination of pK_a^* indicates that the molecule is a stronger acid in the excited state than in the ground state as is expected for a molecule in which the optical electron is nonbonding. However, the shift of the long wavelength π,π^* band of 3,4-benzocinnoline on protonation indicates that the molecule is a weaker acid in the excited state. Moreover, in the conjugate acid, the lowest excited singlet state is π,π^*. Thus if the natures of fluorescing states or the presence of hidden absorption bands are ignored, correlations of Förster cycle data with fluorimetric titration data may be invalid.

5. pK_a^* As an Index of Chemical Reactivity and Structure in the Lowest Excited Singlet State

Phenolic groups, aromatic amines, and sulfhydryl groups are electron donors and generally tend to become more acidic in the lowest excited singlet state relative to the ground state.[56] Carbonyl groups, carboxyl groups, and nitrogen atoms in six-membered heterocycle rings are electron acceptors and tend to become more basic in the lowest excited singlet state than in the ground state.[56] Exceptions to the latter rule,

[67] R. E. Ballard and J. W. Edwards, *Spectrochim. Acta* **20**, 1275 (1964).

however, are to be found in the cases of 3,4-benzocinnoline[67] where the nitrogen atom becomes more acidic in the excited n,π^* state and in doubly protonated o-phenanthroline where solvent relaxation overwhelms charge transfer effects in the π,π^* excited state. Excited state prototropism of five-membered heterocycles has been studied by Bridges and Williams,[68] Chen,[69] and Capomacchia and Schulman.[10] In alkaline solutions, indole and carbazole lose a proton in the excited state but not in the ground state. This shows that a pyrrolic type nitrogen behaves like a secondary aromatic amine and becomes more acidic in the lowest excited singlet state.[70-72] The quenchings of the fluorescences of indole and carbazole in acid solutions are undoubtedly caused by protonation in the excited state.[10,68] However, the site of protonation of indole in the ground state is the 3-carbon atom and not the nitrogen atom.[73] No doubt the basicity of pyrrolic nitrogen is diminished in the lowest excited singlet state. Thus protonation of indole and carbazole in the excited state do not occur at pyrrolic nitrogen. In a study of riboflavin and several structurally related isoalloxazine derivatives, Schulman[74] was able to show, by comparison of the fluorimetrically determined pK_a^* values with those obtained from Förster cycle calculations employing absorption maxima alone, that the equilibria occurring in the flavins corresponded to protonation at different nitrogen atoms in the isoalloxazine ring for ground and excited state reactions, respectively.

Wehry,[75] has shown by means of Förster cycle calculations that the enhancement of the acidity of excited phenols substituted with groups containing sulfur atoms bearing a formal positive charge is caused by valence shell expansion of sulfur in the lowest excited singlet state. In this process the vacant $3d$ orbitals of sulfur accept charge from the aromatic system and thereby exert an acidity strengthening effect. Schulman has shown, by fluorimetric titration, that the anomalously low pK_a^* value for the lowest excited singlet state equilibrium between the cation and zwitterion derived from 8-mercaptoquinoline, and the anomalously short emission wavelength of the 8-mercaptoquinoline cation relative to that of 8-hydroxyquinoline could be explained by

[68] J. W. Bridges and R. T. Williams, *Biochem. J.* **107**, 225 (1968).
[69] R. F. Chen, *Proc. Nat. Acad. Sci. U.S.* **60**, 598 (1968).
[70] H. Boaz and G. K. Rollefson, *J. Amer. Chem. Soc.* **72**, 3435 (1950).
[71] A. Weisstuch and A. C. Testa, *J. Phys. Chem.* **72**, 1982 (1968).
[72] S. G. Schulman, A. C. Capomacchia, and M. S. Rietta, *Anal. Chim. Acta* **56**, 91, (1971).
[73] A. Albert, this treatise Vol. 1, p. 48.
[74] S. G. Schulman, *J. Pharm. Sci.* **60**, 628 (1971).
[75] E. L. Wehry, *J. Amer. Chem. Soc.* **89**, 41 (1967).

valence shell expansion of the sulfur atom of the protonated 8-mer-captoquinoline, in the lowest excited singlet state.[76] Moreover, since the shifts in the absorption spectra of 8-mercaptoquinoline on protona-tion are similar to the shifts in the absorption spectra of 8-hydroxy-quinoline on protonation, valence shell expansion of sulfur must occur subsequent to excitation.

The protonated pyridinic nitrogen atom of the excited 8-mercap-toquinolium ion is capable of inducing valence shell expansion in a sulfur atom which does not normally have the formal positive charge required for $d\pi \leftarrow p\pi$ interaction. This indicates that the inductive effect of the protonated nitrogen atom is much greater in the excited state than in the ground state and is supported by the large difference between the pK_a and pK_a^* (15 units) of the 8-hydroxyquinolinium ion in equilibrium with its zwitterion. Ordinary phenolic dissociations (e.g., α-naphthol) are usually no more than 10 units different in ground and excited states.

C. Phosphorescence and the Acid–Base Chemistry of the Lowest Triplet State

Because of the long lifetimes of molecules in electronically excited triplet states, nonradiative vibrational and collisional processes compete effectively with phosphorescence for deactivation of the lowest excited triplet state. It is this very fact that makes the photochemistry of the triplet state so much more diverse than that of the lowest excited singlet state. However, effective competition of nonradiative deactiva-tional processes with phosphorescence results in phosphorescence in fluid solutions being extremely rare. As a result, phosphorimetry is usually practiced in rigid solutions at liquid nitrogen temperatures where diffusion-regulated processes do not affect the phosphorescence. Moreover, at liquid nitrogen temperature, thermal energy is so low that even unimolecular reactions do not normally follow excitation. The states of ionization of phosphorescing molecules are determined solely by the ground state pK_a and the pH of the test solution prior to freezing.

In spite of the relatively uninteresting dependence of phosphores-cence on pH in rigid solutions, some useful information concerning the chemistry of the lowest excited triplet state is to be had from the study of the shifts of phosphorescence occurring as a result of protonation and dissociation. The Förster cycle [Eq. (20)] may be applied to pH-dependent shifts of phosphorescence spectra in the same way it is

[76] S. G. Schulman, *Anal. Chem.* **44**, 400 (1972).

applied to shifts or absorption of fluorescence spectra. Thus if the pK_a for a particular prototropic equilibrium is known and $\bar{\nu}_A$ and $\bar{\nu}_B$ are taken to be the O–O bands of the phosphorescence spectra of acid and conjugate base, respectively, the dissociation constant (pK_T*) for the acid in the lowest excited triplet state can be calculated. Because phosphorescence is measured in rigid solutions where proton transfer and thus phototautomerism cannot occur subsequent to excitation, the pK_T* will always correspond to the same equilibrium to which pK_a corresponds. At low temperatures, vibrational resolution of spectra, not apparent in fluid solutions at room temperature, is more obvious owing to the elimination of "hot bands" resulting from the thermal populations of higher vibrational levels of ground and excited states, and dynamic solvent phenomena. Thus the O–O band of phosphorescence can usually be identified or at least approximated from the position of the shortest wavelength vibronic maximum in the phosphorescence spectrum[34] and errors in the estimation of pK_T* owing to the inclusion of vibrational energy are virtually nonexistent. This is a fortunate circumstance since the averaging of absorption and phosphorescence spectra in this case would require the measurement of singlet–triplet absorption spectra which are, in most cases, impossible to obtain owing to the extremely low probability of direct singlet–triplet electronic processes.

An alternative but more involved method for the determination of pK_T* values is available in the form of flash photolysis, to produce an appreciable number of triplet molecules, followed by rapid determination of the pH dependence of the triplet–triplet absorption spectrum. The results obtained from the flash photolysis method for several aromatic phenols, carboxylic acids, amines, and heterocycles have been in excellent agreement with those obtained from phosphorescence shifts employed in conjunction with the Förster cycle.[77] Because the triplet–triplet absorption spectra are taken in fluid solutions containing triplet molecules which are fully relaxed with respect to both vibrational manifold and solvent cage, the agreement between the phosphorescence and absorption data indicate that vibrational errors are indeed insignificant in the phosphorescence shift approach and that solvent relaxation is not important as a factor in determining the value of pK_T*. The latter result is a reasonable one because the smaller degree of electronic repulsion in the triplet state results in smaller electronic polarization differences and thus smaller solvent configurational differences than in

---

[77] G. Jackson and G. Porter, *Proc. Roy. Soc. London* **A260**, 13 (1961).

the lowest excited singlet state (where electronic repulsions are much greater relative to the ground state). As a result of the latter circumstances, differences between ground and triplet state dissociation constants tend to be smaller than differences between ground and lowest excited singlet state pK_a values. For example, the two triplet state dissociation constants of o-phenanthroline[78] are intermediate in their values between the corresponding ground and lowest excited singlet state dissociation constants.[16]

Schulman et al.[79] studied the effect of the nitro group on the dissociations of several nitrophenols and nitroquinolines in ground, lowest excited singlet and lowest excited triplet states. It was found that in the lowest triplet state the acidity enhancing effect of the nitro group parallels that in the ground state. However, in the lowest excited singlet state, the enhancement of acidity by the nitro group is much more exaggerated than in the ground state owing to the charge transfer acceptor nature of the nitro group in the lowest excited singlet state.

Finally, it is obvious that the lowest excited singlet state is much more different from the ground state than is the lowest excited triplet state. It is probably a safe assumption that if the excited singlet state was longer lived, the photochemistry that would result would be of much greater diversity than the photochemistry of the lowest triplet state.

ACKNOWLEDGMENT

The author is grateful to Mrs. Sandra Williams, Mrs. Meriam Stallings, and Mr. Anthony Capomacchia for assistance with the preparation of this manuscript.

---

[78] J. S. Brinen, D. D. Rosebrook, and R. C. Hirt, *J. Amer. Chem. Soc.* **67**, 2651 (1963).
[79] S. G. Schulman, L. B. Sanders, and J. D. Winefordner, *Photochem. Photobiol.* **13**, 381 (1971).

197

Some Applications of Thermochemistry
to Heterocyclic Chemistry

KALEVI PIHLAJA AND ESKO TASKINEN

DEPARTMENT OF CHEMISTRY, UNIVERSITY OF TURKU, 20500 TURKU 50, FINLAND

I. INTRODUCTION

The scope of the present article is so wide that it is necessary to limit the discussion to certain selected topics, especially since several recent reviews deal with some aspects of heats of combustion, heats of formation, heats of hydrogenation, and bond energies.[1-3] The main aim of this account is to consider results which have not previously been reviewed. Applications of enthalpies of formation and energy parameters obtainable by chemical equilibration are of considerable significance. They are useful in the determination of reaction heats, resonance energies, ring strains, conformational, and other nonbonded energy entities and are therefore considered in some detail.

Throughout, the emphasis is on the chemical knowledge obtained by applications of the different thermochemical methods rather than on the principles of these methods. The reader interested in experimental procedures is provided with suitable citations in the text.

The first part of the following discussion considers new enthalpy of formation data and the use of an Allen-type group increment scheme for evaluation of heats of formation, nonbonded interactions, and conformational energies. The second part concentrates on the determination of different energy parameters with the aid of chemical equilibration, and the last two sections concentrate on the concepts of ring strain and resonance energy in heterocyclic compounds.

II. ENTHALPIES OF FORMATION

A. New Experimental Values

Since the comprehensive compilation by Pilcher and Cox[1] was published in 1970, heats of formation of various additional heterocyclic compounds have become available. Good[4] measured the enthalpies of formation for nine organic nitrogen compounds (1-9; Table I) and also for five sulfur-containing heterocycles[5] (10-14; Table I). Månsson et al.[6] reported the heats of formation of trioxane [15] and tetroxane [16] and Pihlaja and Luoma[7] for 1,3-dioxane [17] and thirteen methyl derivatives (Table I). Enthalpies of formation of 1,3-dioxolane [18] and

---

[1] J. D. Cox and G. Pilcher, "Thermochemistry of Organic and Organometallic Compounds." Academic Press, New York, 1970.

[2] R. T. Sanderson, "Chemical Bonds and Bond Energy." Academic Press, New York, 1971.

[3] C. T. Mortimer, "Reaction Heats and Bond Strengths." Pergamon, Oxford, 1962.

[4] W. D. Good, J. Chem. Eng. Data 17, 28 (1972).

[5] W. D. Good, J. Chem. Eng. Data 17, 158 (1972).

[6] M. Månsson, E. Morawetz, Y. Nakase, and S. Sunner, Acta Chem. Scand. 23, 56 (1969).

[7] K. Pihlaja and S. Luoma, Acta Chem. Scand. 22, 2401 (1968).

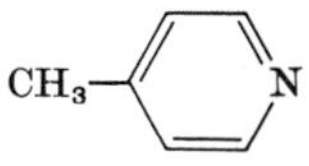

[1] 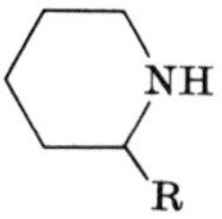   [2] R = H   [4]   [5]
       [3] R = CH$_3$

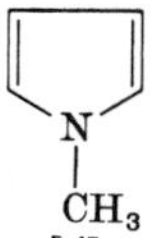

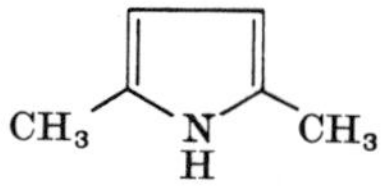

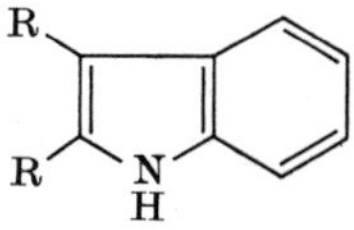

[6] R = H   [8]   [9]
[7] R = CH$_3$

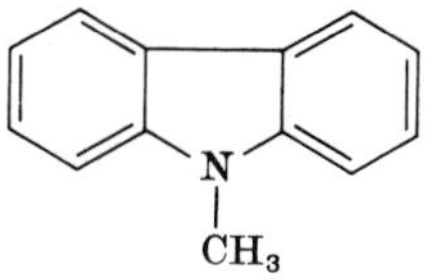

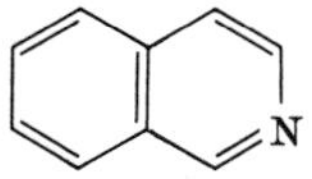

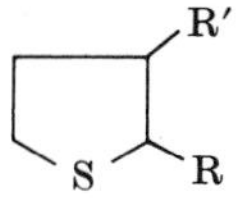

[10] R = CH$_3$,   [12] 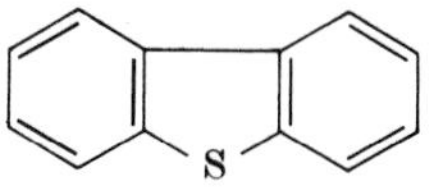   [13] 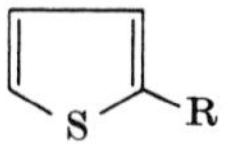   [14]
     R' = H   R = i-Pr
[11] R = H,
     R' = CH$_3$

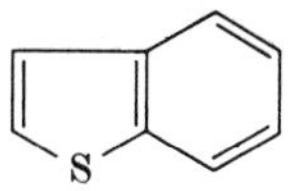

two methyl derivatives have been measured[8,9] as well as those of 2-(2-methoxyethoxy)tetrahydropyran [19][10] and 5-methyl-2,3-dihydrofuran [20][11] (Table I). Choi and Joncich[12] gave enthalpies of formation for

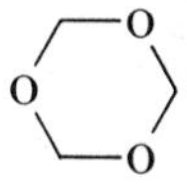

[15] 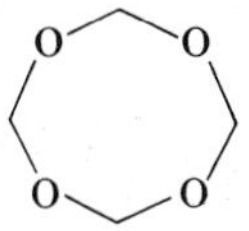   [16] 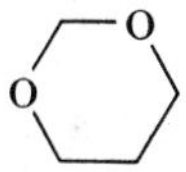   [17]

some cyclic carbonates but their experimental details are very poorly described and the obtained results seem to be so inaccurate that they must be considered tentative at best. Table I also contains some older values of heats of formation employed later in comparisons.

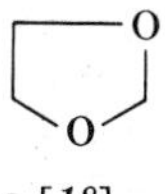

[18] 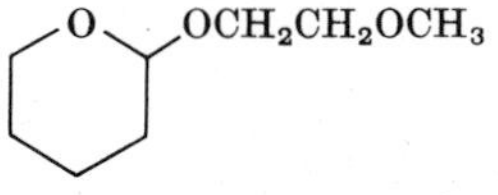   [19] 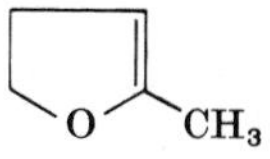   [20]

[8] K. Pihlaja and J. Heikkilä, *Acta Chem. Scand.* **23**, 1053 (1969).
[9] K. Pihlaja and J. Heikkilä, *Suom. Kemistilehti B* **45**, 148 (1972).
[10] K. Pihlaja and M.-L. Tuomi, *Acta Chem. Scand.* **24**, 366 (1970).
[11] K. Pihlaja and J. Heikkilä, *Suom. Kemistilehti B* **42**, 338 (1969).
[12] J. K. Choi and M. J. Joncich, *J. Chem. Eng. Data* **16**, 87 (1971).

TABLE I

EXPERIMENTALLY DETERMINED AND CALCULATED ENTHALPIES OF FORMATION FOR SOME HETEROCYCLIC COMPOUNDS

Compound	$-\Delta H_f^\circ$(l or c) (kJ mole^{-1})	$-\Delta H_f^\circ$(g) (kJ mole^{-1})		C.R.S.E.^{a}	Ref.
		Exptl.	Calc.		
4-Methylpyridine [1]	−59.2 ± 0.9	−103.8			g
Piperidine [2]	86.4 ± 0.6	47.2	48.7	1.5	g
2-Methylpiperidine [3]	88.5 ± 1.1	49.2	48.7	−0.5	h
2-Methylpyrrole [4]	124.9 ± 1.1	84.5	82.4	−2.1	g
1-Methylpyrrole [4]	−62.4 ± 0.5	−103.1			g
2,5-Dimethylpyrrole [5]	16.8 ± 0.8	−39.75			g
Indole [6]	−86.65 ± 0.75^b				g
2,3-Dimethylindole [7]	−4.2 ± 1.0^b				g
9-Methylcarbazole [8]	−105.5 ± 1.1^b				g
Isoquinoline [9]	−145.1 ± 0.9				i (h)
2-Methylthiacyclopentane [10]	105.1 ± 0.7	63.9 (63.3)	71.1	7.2 (7.8)	i
3-Methylthiacyclopentane [11]	102.3 ± 0.7	60.2	67.7	7.5	i
2,3-Benzothiophene [12]	−100.9 ± 0.9^b				i
Dibenzothiophene [13]	−120.3 ± 1.5^b				i
2-Isopropylthiophene [14]	11.6 ± 1.8				i
Trioxane [15]	522.3 ± 0.4^b	465.8	492.0	26.2	j
Tetroxane [16]	699.9 ± 0.5^b	620.2	656.0	35.8	j
Trioxane [15]	520.4	463.8	492.0	28.2	k,l
1,3-Dioxane [17]	384.6 ± 0.9	349.0	350.9	1.9	m
	386.2	350.6	350.9	0.3	h
1,3-Dioxolane [18]	337.2 ± 0.7	301.7	321.7	20.0	n
2-Methyl-1,3-dioxolane	386.9 ± 0.9	352.0	366.0	14.0	n
	385.05 ± 1.1	350.2	366.0	15.8	o
2,4-Dimethyl-1,3-dioxolane	420.8 ± 1.2	381.9	404.0	22.1	n
2,2-Dimethyl-1,3-dioxolane	423.0 ± 2.9	386.4	405.6	19.2	h
2-(2-Methoxyethoxy)tetrahydropyran [19]	623.3 ± 2.1	563.2	561.4	−1.8	p
5-Methyl-2,3-dihydrofuran [20]	165.0 ± 1.1	130.2			q
3,6-Dioxaoctane [21]	451.4 ± 1.0	408.2	416.1	7.9^c	r
2,4,6-Trimethyl-1,3,5-trioxane [23]	681.8 ± 1.3	640.4	623.9	−16.5	s
		645.2	623.9	−21.3	s

202

Compound					
1,4-Dioxane [22]	353.4 ± 0.6	315.9	316.1	0.2	h
4-Methyl-1,3-dioxane	436.5 ± 1.3	387.3	389.2	1.9	t
2-Methyl-1,3-dioxane	474.8 ± 1.2	396.6[d]	394.8	−1.8	m,u
cis-4,6-Dimethyl-1,3-dioxane	462.6 ± 0.9	432.2[d]	427.7	−4.5	m
trans-4,6-Dimethyl-1,3-dioxane	465.2 ± 2.1	418.1[d]	e		m
cis-2,4-Dimethyl-1,3-dioxane	468.9 ± 1.0	424.0[d]	433.2	9.2	m,u
2,2-Dimethyl-1,3-dioxane	461.3 ± 1.1	426.4[d]	e		m
5,5-Dimethyl-1,3-dioxane	451.7 ± 1.0	418.7[d]	e		m,u
4,5-Dimethyl-1,3-dioxane (cis + trans 1:2)		407.5[d]	e		m,u
2,2,4-Trimethyl-1,3-dioxane	500.9 ± 2.1	457.3[d]	e		m,u
4,4,6-Trimethyl-1,3-dioxane	500.6 ± 1.7	455.1[d]	e		m
r-2-cis-4-trans-6-Trimethyl-1,3-dioxane	488.3 ± 1.4	443.6[d]	e		m,u,v
2,2-cis-4,6-Tetramethyl-1,3-dioxane	539.4 ± 1.7	494.8[d]	e		m,u
2,2-trans-4,6-Tetramethyl-1,3-dioxane	526.3 ± 2.4	482.7[d]	f		m,u

[a] See Section II,B,2.

[b] These values refer to the crystalline state.

[c] Δ_{occo} (Section II,B,1).

[d] For heats of vaporization see I. Wadsö, Acta Chem. Scand. 20, 544 (1966).

[e] These compounds include various nonbonded interactions due to axial methyl groups.

[f] This compound exists predominantly in the twist-boat conformation (Section II,C).

[g] W. D. Good, J. Chem. Eng. Data 17, 28 (1972).

[h] J. D. Cox and G. Pilcher, "Thermochemistry of Organic and Organometallic Compounds." Academic Press, New York, 1970.

[i] W. D. Good, J. Chem. Eng. Data 17, 158 (1972).

[j] M. Månsson, E. Morawetz, Y. Nakase, and S. Sunner, Acta Chem. Scand. 23, 56 (1969).

[k] M. Delepine and M. Badoche, C. R. Acad. Sci. Paris 214, 777 (1942).

[l] W. K. Busfield and D. Merigold, J. Chem. Soc. A 2975 (1969); Makromol. Chem. 138, 65 (1970).

[m] K. Pihlaja and S. Luoma, Acta Chem. Scand. 22, 2401 (1968).

[n] K. Pihlaja and J. Heikkilä, Acta Chem. Scand. 23, 1053 (1969).

[o] K. Pihlaja and J. Heikkilä, Suom. Kemistilehti B 45, 148 (1972).

[p] K. Pihlaja and M.-L. Tuomi, Acta Chem. Scand. 24, 366 (1970).

[q] K. Pihlaja and J. Heikkilä, Suom. Kemistilehti B 42, 338 (1969).

[r] M. Månsson, J. Chem. Thermodynam. 1, 141 (1969).

[s] K. Pihlaja and M.-L. Tuomi, Suom. Kemistilehti B 43, 224 (1970).

[t] K. Pihlaja and J. Kankare, Acta Chem. Scand. 23, 1745 (1969); K. Pihlaja, ibid. 25, 451 (1971).

[u] K. Pihlaja and J. Heikkilä, Acta Chem. Scand. 21, 2390, 2430 (1967).

[v] For the new nomenclature, see J. Org. Chem. 35, 2849 (1970).

B. Allen-Type Group Increments for Saturated Compounds Containing Oxygen, Sulfur, or Nitrogen

1. General Treatment

The background to this type of theoretical application of thermochemical data can best be obtained from Pilcher and Cox[1] and from references cited therein. Kalb, Chung, and Allen[13] computed a group increment scheme (for alkanes) involving six or seven parameters. A similar scheme was later compiled for certain organic oxygen compounds.[14a,b] Because of the similarity of corresponding increments in both schemes, they have been combined[15] by fitting data for 37 saturated oxygen-containing compounds and 54 alkanes to Eq. (1)[13,14a,b]

$$\Delta H_a{}^0(g) - \Sigma E_b(g) = n_1\Gamma_{CCC} + n_2\Gamma_{CCO} + n_3\Gamma_{COC} + n_4\Gamma_{OCO}$$
$$+ n_5\Delta_{CCC} + n_6\Delta_{CCO} + n_7\Delta_{OCO} + n_8 S_{15}{}^{HH}$$
$$+ n_9 S_{16}{}^{HH}(C) + n_{10} S_{16}{}^{HH}(O) + n_{11} S_L \qquad (1)$$

where $\Delta H_a{}^0(g)$ is the heat of atomization of the compound in question[1,14a,b,15] and $\Sigma E_b(g)$ the sum of the respective bond energies in the gaseous state (Table II). The structural parameters n_1, n_2, etc., correspond to the different interactions in the molecule. The general term Γ_{ABC} illustrates the effect of the interaction ABC and Δ_{ABC} means the trigonal interaction due to three nonbonded atoms ABC attached to the same carbon atom. $S_{15}{}^{HH}$ is the contribution of the gauche-n-butane type nonbonded interaction between fifth-neighbor hydrogen atoms; $S_{16}{}^{HH}$(C or O) are the contributions of the nonbonded interactions between sixth-neighbor hydrogen atoms;[14a,b] and S_L is the so-called locking parameter in alkanes.[13] The value of $S_L(0)$ and $S_{15}{}^{HO}$ were shown to be negligible.[15] The values of the enthalpies of formation of the gaseous atoms and those of the bond energies in the gaseous state used for the left-hand side of Eq. (1) are shown in Table II and the values of the group increments computed using a linear regression program with a least-squares criterion in Table III.[14a,b,15]

The values derived as described above for the common group increments and the enthalpies of formation of 31 gaseous saturated sulfur-containing compounds were used to obtain the values of Γ_{CCS}, Γ_{CSC},

[13] A. J. Kalb, A. L. H. Chung, and T. L. Allen, J. Amer. Chem. Soc. 88, 2938 (1966).
[14a] K. Pihlaja and J. Kankare, Acta Chem. Scand. 23, 1745 (1969).
[14b] K. Pihlaja, Acta Chem. Scand. 25, 451 (1971).
[15] K. Pihlaja and P. Vainiotalo, private communication (1972).

TABLE II

ENTHALPIES OF FORMATION OF THE GASEOUS ATOMS AND THE BOND ENERGIES IN THE
GASEOUS STATE USED IN THE EVALUATION OF THE GROUP INCREMENTS IN TABLE III

Atom	$-\Delta H_f{}^0(g)$ (kJ mole^{-1})
C	715.0$_5$
H	218.0
O	249.2
S	274.7
N	472.8

Bond	$E_b(g)$ (kJ mole^{-1})
C—C	329.9
C—H	415.4
C—O	327.0
O—H	463.5
C—S	272.6
S—H	365.5
C—N	272.5$_5$
N—H	391.0
N—N	158.4

and Δ_{CCS} using the same computer program.[14a,b,15] Similarly, the enthalpies of formation of 14 gaseous saturated nitrogen-containing compounds allowed the evaluation of Γ_{CCN}, Γ_{CNC}, Γ_{CNN}, Δ_{CCN}, Δ_{CCC}^N, and Δ_{CCN}^N increments.[15] The superscript in the last two increments means that nitrogen is the central atom to which the three subscript atoms are attached. These results are also presented in Table III.

These group increments allow calculation of further increments if reasonably accurate data are available. For instance, Månsson determined the heat of formation of 3,6-dioxaoctane [21] to be -408.2 kJ

$$CH_3CH_2OCH_2CH_2OCH_2CH_3$$
[21]

mole^{-1}.[16] Its heat of atomization is then 8248.7 kJ mole^{-1} and the sum of the bond energies 8113.4 kJ mole^{-1}. The left-hand side of Eq. (1) is then 135.3 kJ mole^{-1}. Using the appropriate group increments (Table III), the right-hand side of Eq. (1) is given as 143.2 kJ mole^{-1} which indicates that -7.9 kJ mole^{-1} is the interaction between 1,4-oxygen atoms (Δ_{OCCO}).

[16] M. Månsson, *J. Chem. Thermodynam.* **1**, 141 (1969).

TABLE III

OPTIMUM VALUES OF SOME GROUP INCREMENTS CALCULATED FOR SATURATED OXYGEN, SULFUR, OR NITROGEN COMPOUNDS[c]

Increment	Contribution to $-\Delta H_f^0(g)$ (kJ mole^{-1})
Γ_{CCC}	11.23 ± 0.05[a]
Γ_{CCO}	23.62 ± 0.28
Γ_{COC}	24.40 ± 0.61
Γ_{OCO}	54.96 ± 0.78
Γ_{CCS}	13.87 ± 0.13
Γ_{CSC}	11.81 ± 0.26
Γ_{CCN}	17.46 ± 0.50
Γ_{CNC}	16.33 ± 0.65
Γ_{CNN}	24.64 ± 0.73
Δ_{CCC}	-3.05 ± 0.19
Δ_{CCO}	-6.12 ± 0.39
Δ_{OCO}	$-12.94 + 0.72$
Δ_{CCS}	$-4.94_5 \pm 0.18$
Δ_{CCN}	-4.70 ± 0.77
Δ_{CCC}^{N}[b]	-6.66 ± 1.83
Δ_{CCN}^{N}[b]	-7.79 ± 2.28
S_{15}^{HH}	-2.66 ± 0.17
$S_{16}^{HH}(C)$	-12.59 ± 1.68
$S_{16}^{HH}(O)$	-6.72 ± 1.81
$S_L(C)$	-1.44 ± 0.25

[a] Standard deviations.
[b] The superscript means that N is the central atom in these nonbonded trios.
[c] A. J. Kalb, A. L. H. Chung, and T. L. Allen, *J. Amer. Chem. Soc.* **88**, 2938 (1966); K. Pihlaja and J. Kankare, *Acta Chem. Scand.* **23**, 1745 (1969); K. Pihlaja, *Acta Chem. Scand.* **25**, 451 (1971); K. Pihlaja and P. Vainiotalo, Private communication, (1972).

The following describes the use of such increments (Table III) to estimate nonbonded interactions and ring strain.

2. Evaluation of Ring Strain for Some Saturated Heterocycles

Section IV deals in more detail with the concept of ring strain (conventional ring strain energy, or C.R.S.E.[1]), but we will now show briefly the use of the group increments in Table III for the estimation of ring strain. For nonsubstituted monocyclic rings, Eq. (2) holds by definition.[1]

$$C.R.S.E. = \Delta H_f^0(g)_{exptl.} - \Delta H_f^0(g)_{calcd.} \qquad (2)$$

Equation (1) relates the calculated heats of formation to the heats of atomization.

The calculated heats of formation for the gaseous 2-methyl- [*10*] and 3-methylthiacyclopentanes [*11*] are -71.1 and -67.7 kJ mole^{-1}, respectively, assuming that the latter compound includes one S_{15}^{HH} interaction.[17] Comparison with the experimental values shows a ring strain of 7.5 kJ mole^{-1} in both cases in excellent agreement with the C.R.S.E. value obtained for the parent compound (7.6 kJ mole^{-1}).[1]

Piperidine [*2*] and 2-methylpiperidine [*3*] are almost strain free (C.R.S.E.s -0.5 and -2.1 kJ mole^{-1}, respectively) as are most other saturated six-membered heterocycles (Section IV).

The calculated enthalpy of formation of 1,3-dioxolane [*18*] is -322.0 kJ mole^{-1}, which means the C.R.S.E. is about 20.0 kJ mole^{-1}, in close agreement with the average strain in the methyl-substituted derivatives (19.0 kJ mole^{-1}) shown in Table I. The contribution, -7.9 kJ mole^{-1}, of the 1,4-oxygen atoms has been taken into account when calculating the ring strain of 1,3-dioxolanes.

The ring strain of 1,4-dioxane [*22*] includes two interactions (Δ_{OCCO}) due to the 1,4-oxygen atoms. As the calculated enthalpy of formation, -316.1 kJ mole^{-1}, is practically equal to the experimental value of -315.9 kJ mole^{-1}, this six-membered ring is also strain free.

[*22*]

Månsson *et al.*[6] reported the enthalpy of formation of trioxane (15) as -465.8 kJ mole^{-1}, in close agreement with the older estimate of -463.8 kJ mole^{-1} by Delepine and Badoche.[18] Moreover, Busfield and Merigold[19a,b] measured the enthalpy change for the reaction

$$3 \ CH_2O \ (g) \longrightarrow \text{Trioxane (s)}$$

as -194.6 kJ mole^{-1}. The heat of formation of gaseous formaldehyde, -108.6 kJ mole^{-1},[20,21] gives -520.4 kJ mole^{-1} for the enthalpy of formation of solid trioxane which implies a value of -463.8 kJ mole^{-1} in the gaseous state.[6] The close agreement between these three separate results makes them preferable to the values based on the work of Walker and Carlisle[22] generally used so far.[14a,b]

[17] R. Keskinen and K. Pihlaja, private communication dealing with the conformational preferences in five-membered heterocycles (1972).

[18] M. Delepine and M. Badoche, *R. Acad. Sci. Paris* **214**, 777 (1942).

[19a] W. K. Busfield and D. Merigold, *J. Chem. Soc. A* 2975 (1969).

[19b] W. K. Busfield and D. Merigold, *Makromol. Chem.* **138**, 65 (1970).

[20] G. I. Birley and H. A. Skinner, *Trans. Faraday Soc.* **66**, 791 (1970).

[21] R. A. Fletcher and G. Pilcher, *Trans. Faraday Soc.* **66**, 794 (1970).

[22] J. F. Walker and P. J. Carlisle, *Chem. Eng. News* **21**, 1250 (1943).

Although trioxane has appreciable ring strain (27.5 kJ mole^{-1}), 2,4,6-trimethyl-1,3,5-trioxane (paraldehyde [23]) shows a stabilization effect of 16.5–21.3 kJ mole^{-1}. The possible basis of this phenomenon has been discussed in detail elsewhere.[14a,b,23]

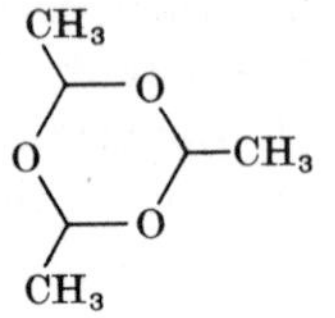

[23]

C. Estimation of Conformational Interactions with the Aid of Enthalpies of Formation

Pihlaja et al.[7,24] determined heats of formation for several methyl-substituted 1,3-dioxanes in the liquid state, and evaluated the contributions of equatorial and axial methyl groups in the different ring positions. The differences in the derived contributions were shown to be equal to the conformational energies of axial methyl groups in the different positions of the ring and the values obtained were in close agreement with those from chemical equilibration (Table VII in Section III).

The enthalpies of formation of the isomeric 2,2,4,6-tetramethyl-1,3-dioxanes[7,25] in the liquid state differ by 13.1 kJ mole^{-1} in favor of the 4,6-cis isomer [24] which exists predominantly in the chair form. The 4,6-trans isomer assumes the twist-boat form [25a] to avoid the strong 2,6-syn-axial methyl–methyl interaction present in the chair form [25b]. The 4,6-cis isomer has an axial 2-methyl group with an interaction energy of 17.0 kJ mole^{-1},[7,26] enhanced in the liquid state by 3.1 kJ mole^{-1}.[27] The chair conformation is therefore 13.1 + 17.0 + 3.1 = 33.2 kJ mole^{-1} more stable than the twist boat.

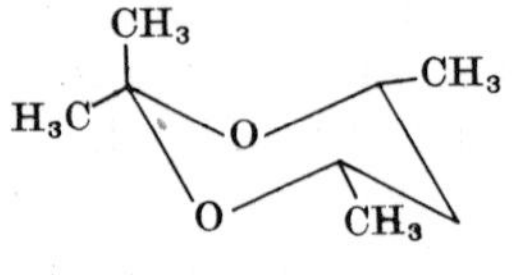

[24] 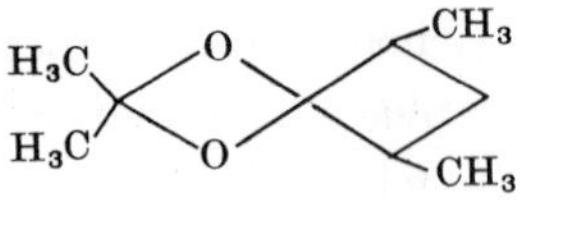   [25a] 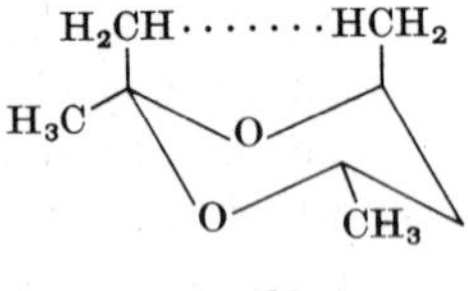   [25b]

23 K. Pihlaja and M.-L. Tuomi, Suom. Kemistilehti B 43, 224 (1970).
24 K. Pihlaja and J. Heikkilä, Acta Chem. Scand. 21, 2390, 2430 (1967).
25 K. Pihlaja, Acta Chem. Scand. 22. 716 (1968).
26 F. W. Nader and E. L. Eliel, J. Amer. Chem. Soc. 92, 3050 (1970).
27 E. L. Eliel, J. R. Powers, Jr., and F. W. Nader, private communication (1971). See also refs. 54b and c.

It is also possible to use the group increments to find the enthalpy change for the chair–twist interconversion (ΔH_{ct}) for the 1,3-dioxane ring in the gaseous state. The calculated heat of formation of the hypothetical strainless chair form of 2,2-*trans*-4,6-tetramethyl-1,3-dioxane is -514.6 kJ mole^{-1}. The enthalpy of formation of 2,2-*cis*-4,6-tetramethyl-1,3-dioxane without the contribution of the interaction due to the axial 2-methyl group is also -514.6 kJ mole^{-1}. The experimental values in the gaseous state are -482.7 kJ mole^{-1} for the *trans* isomer and -494.8 kJ mole^{-1} for the *cis* isomer. Hence the interaction energy of the axial 2-methyl group in the *cis* isomer is about 19.8 kJ mole^{-1} in the gaseous state ($17.0 + 3.1 = 20.1$ kJ mole^{-1} in the liquid state; see above) and the value of $\Delta H_{ct}(g)$ is about 31.9 kJ mole^{-1}, close to that estimated above for the liquid state.[28,29]

Slightly deviating values are obtained for the calculated enthalpies of formation from the recent combined scheme[15] and from the separate schemes presented earlier for alkanes[13] and their oxygen homologs.[14a,b] However, the combined scheme decreases appreciably the number of different increments, leaves the main conclusions intact, does not effect the general accuracy,[15] and hence is preferred.

III. CHEMICAL EQUILIBRATION

A. Introduction

Chemical equilibration is one of the most accurate and important tools to measure conformational and other energy differences between stereoisomers or other structural isomers. Methods depend on the problem to be solved.[30] We deal mainly with epimerizations of saturated heterocycles of the type [26] and [27], and also with equilibrations of *endo-exo* isomeric cyclic vinyl ethers like [28] and [29] in the presence of acid catalysts.

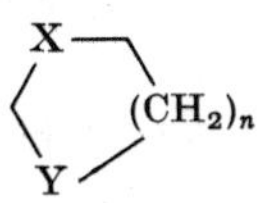

[26] 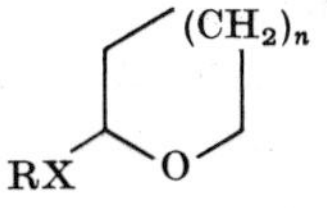   [27] 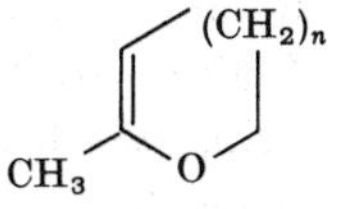   [28] 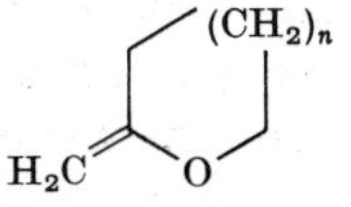   [29]

[28] K. Pihlaja and J. Jalonen, *Org. Mass Spectrom.* **5**, 1363 (1971).
[29] G. M. Kellie and F. G. Riddell, "Non-Chair Conformations of Six-Membered Rings," *in* "Topics in Stereo-chemistry" (N. L. Allinger and E. L. Eliel, eds.), Vol. 7. Wiley (Interscience), New York, 1973.
[30] E. L. Eliel, N. L. Allinger, S. J. Angyal, and G. A. Morrison, "Conformational Analysis," pp. 58–70, 141–142. Wiley, New York, 1965.

B. Five-Membered Heterocycles

1. *General remarks*

Conformational analysis of five-membered heterocycles has progressed more slowly than that of the corresponding six-membered analogs. Now that the "envelope" [*30*] or "half-chair" [*31*] forms are recognized as inadequate models for a description of the conformations of many five-membered ring systems, progress has occurred.[31,32] The older approach was oversimplified; many five-membered rings possess several minimum-energy conformations between [*30*] and [*31*].

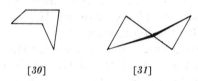

[*30*]   [*31*]

2. *1,3-Dioxolanes*

Willy *et al.*[31] clarified the chemical equilibration of alkyl-substituted 1,3-dioxolanes by NMR spectroscopy, after earlier contradictory reports.[33-36]

In 2,4-dialkyl-1,3-dioxolanes the *cis* form is always thermodynamically more stable; changing the 2-substituent has no effect on the equilibrium, and the effect of the 4-substituent is minor.

The *cis* isomers are stabilized thermochemically; e.g., *cis*- is preferred in entropy to *trans*-2,4-dimethyl-1,3-dioxolane by only 1.1 ± 0.2 J °K^{-1} mole^{-1}.[37]

The *syn* isomers of 2-*r*-4-*cis*-5-trialkyl-1,3-dioxolane are usually thermodynamically more stable.[31] However, 2-*t*-Bu-*r*-4-*cis*-5-di-*i*-Pr- and 2-*r*-4-*cis*-5-tri-*t*-Bu-1,3-dioxolanes exist preferentially in the *anti* forms, probably because of unfavorable 1,3-transannular nonbonded interactions between the bulky substituents. Some equilibration results of alkyl-substituted 1,3-dioxolanes are shown in Table IV.

The five-membered ring of 1,3-dioxolanes is extremely flexible. The

[31] W. E. Willy, G. Binsch, and E. L. Eliel, *J. Amer. Chem. Soc.* **92**, 5394 (1970).
[32] R. Keskinen, A. Nikkilä, and K. Pihlaja, *Tetrahedron* **28**, 3943 (1972).
[33] A. Kankaanperä, Ph.D. Thesis, Univ. of Turku, Turku, Finland (1966).
[34a] M. Anteunis and F. Alderweireldt, *Bull. Soc. Chim. Belg.* **73**, 889 (1964).
[34b] F. Alderweireldt and M. Anteunis, *Bull. Soc. Chim. Belg.* **74**, 488 (1965).
[35a] Y. Rommelaere and M. Anteunis, *Bull. Soc. Chim. Belg.* **79**, 11 (1970).
[35b] G. Lemière and M. Anteunis, *Bull. Soc. Chim. Belg.* **80**, 215 (1971).
[36] For further citations see Willy *et al.*[31] and Keskinen *et al.*[32]
[37] P. Salomaa, *Ann. Univ. Turku, Ser. A* No. 46 (1961).

TABLE IV

EQUILIBRIA IN SOME 2,4-DIALKYL-1,3-DIOXOLANES AND
2,4,5-TRIALKYL-1,3-DIOXOLANES AT 25°C

2-R	4-R	5-R	$-\Delta G^0$ (kJ mole^{-1})
Me	Me	H	1.15[a–c]
i-Pr	Me	H	1.00[a,c]; 1.06[d]; 0.71[e]
t-Bu	Me	H	1.15[a,c]; 1.17[d]
Me	i-Pr	H	1.35[a,c]; 1.38[e]
Me	t-Bu	H	2.05[a,c]
t-Bu	t-Bu	H	1.71[a,c]
Me	Me	Me	2.94[c,f,g]; 3.43[e]
i-Pr	i-Pr	i-Pr	1.78[c,f]
i-Pr	t-Bu	t-Bu	−0.05[c,f]
t-Bu	t-Bu	t-Bu	−3.84[c,f]
Me	(Me)$_2$	Me	1.80[a,e]
Me	Me	H	−0.30[a,h–j]
i-Pr	Me	H	0.45[a,h,i]
t-Bu	Me	H	0.54[a,h,i]

[a] For the *trans* ⇌ *cis* equilibrium.

[b] Reported $\Delta S^0 = 1.1 \pm 0.2$ J K^{-1} mole^{-1} in P. Salomaa, *Ann. Univ. Turku Ser. A* No. 46 (1961).

[c] W. E. Willy, G. Binsch, and E. L. Eliel, *J. Amer. Chem. Soc.* **92**, 5394 (1970).

[d] Y. Rommelaere and M. Anteunis, *Bull. Soc. Chim. Belg.* **73**, 889 (1964).

[e] A. Kankaanpera, Ph.D. Thesis, Univ. of Turku, Turku, Finland (1966).

[f] For the *anti–syn* equilibrium (see text also).

[g] Reported $\Delta S^0 = -1.4 \pm 0.1$ J K^{-1} mole^{-1} in W. E. Willy, G. Binsch, and E. L. Eliel, *J. Amer. Chem. Soc.* **92**, 5394 (1970).

[h] These values refer to 2,4-dialkyl-substituted 1,3-dithiolanes.

[i] R. Keskinen, A. Nikkila, and K. Pihlaja, *J. Chem. Soc. Perkin II* 1376 (1973).

[j] Reported $-\Delta S^0 = 1.0 \pm 0.2$ J K^{-1} mole^{-1} in Keskinen (note i above).

conformational situation in most of the alkyl-substituted 1,3-dioxolanes[31] is fundamentally different from that found in the six-membered rings to be discussed later (Section III,C). Few alkyl-substituted 1,3-dioxolanes clearly possess discrete minimum energy wells in their pseudorotation circuits and a number of them show intermediate behavior.[31]

3. *1,3-Dithiolanes*

Little published data are available on 1,3-dithiolanes [*32*],[38] but the results of Keskinen *et al.*[39] show that the ring is flexible, and only tiny energy differences are observed between the diastereoisomeric 2,4-dialkyl derivatives. Equilibration of the 2,4-dimethyl-1,3-dithiolanes at several temperatures showed that entropy slightly favors the *trans* isomer (by 1.0 ± 0.2 J $°K^{-1}$ mole^{-1}), whereas the enthalpy difference is practically zero (see also Table IV).

[*32*]

In comparison with the correspondingly substituted 1,3-dioxolanes (Section III,B,2), this result reflects the lengths of the C—O bond (142 pm) and the C—S bond (182 pm).

4. *1,3-Oxathiolanes*

NMR studies[40,41] of 1,3-oxathiolanes are incomplete. The results for the symmetric analogs, 1,3-dioxolanes and 1,3-dithiolanes, appear to indicate that 1,3-oxathiolane should be a mobile and rapidly pseudo-rotating system. Careful inspection of models reveals, however, that the different bond lengths (C—O = 142 pm, C—C = 154 pm, and C—S = 182 pm) could by distortion greatly favor some or only one of the possible pseudorotamers.

The early NMR results[40,41] confirmed this in part. Pasto *et al.*[40] calculated values for the conformational energies of the two pseudoaxial alkyl groups from chemical shifts. Keskinen *et al.*[32,42] studied systematically alkyl-substituted 1,3-oxathiolanes by chemical equilibration and NMR.

Epimerization occurs easily (Scheme I); in 1,3-oxathiolanes the ring opening during the epimerization reaction occurs mainly at the C—O bond.[43]

[38] L. A. Sternson, D. A. Coviello, and R. S. Egan, *J. Amer. Chem. Soc.* **93**, 6529 (1971).
[39] R. Keskinen, A. Nikkilä, and K. Pihlaja, *J. Chem. Soc. Perkin II* 1376 (1973).
[40] D. J. Pasto, F. M. Klein, and T. W. Doyle, *J. Amer. Chem. Soc.* **89**, 4368 (1967).
[41] G. E. Wilson, Jr., M. G. Huang, and F. A. Bovey, *J. Amer. Chem. Soc.* **92**, 5907 (1970).
[42] K. Pihlaja, R. Keskinen, and A. Nikkilä, Unpublished results dealing with conformational analysis of 2,4-dialkyl-, 2,2,4-trialkyl, 4,5-dimethyl-, 2,4,5-trimethyl-, and 2,2,4,5-tetramethyl-1,3-oxathiolanes.
[43] K. Pihlaja, *J. Amer. Chem. Soc.* **94**, 3330 (1972).

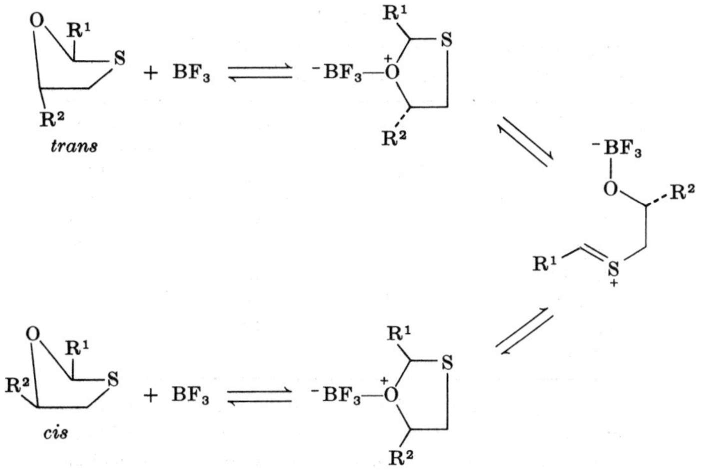

Scheme 1

The equilibrations[32] confirmed the suggestion[40] that the 1,3-oxathiolane ring system is much less mobile than other five-membered ring systems. At least for 5-alkyl-substituted derivatives, equilibria exist between certain conformations [Eq. (3)], and pseudoaxial 2- or 5-alkyl groups possess conformational energies (Table V).[32]

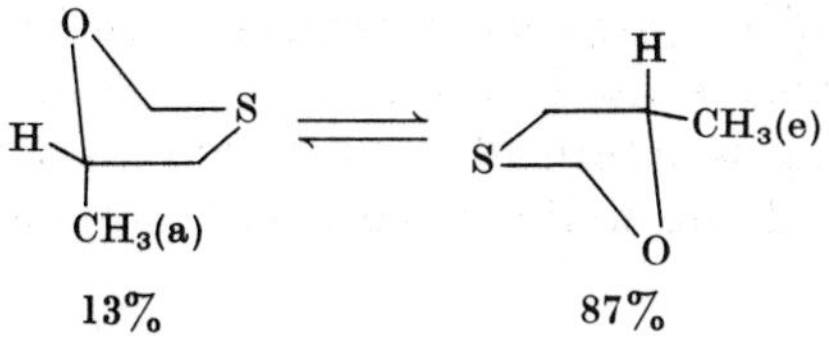

$$\text{(3)}$$

13%   87%

The enhanced 2,5-transannular nonbonded interaction indicates that such 2,4-interaction over the C—S—C moiety might be small and close to that in 1,3-dithiolane derivatives. This was confirmed recently by a conformational study of 4-methyl- and 4,5-dimethyl-1,3-oxathiolanes:[42] cis and trans configurations of 2,4-dialkyl- and 2,2,4-trialkyl-1,3-oxathiolanes are almost equally stable (both ΔH and ΔS are very small),[39,42] whereas, e.g., r-2-cis-4-cis-5-trimethyl-1,3-oxathiolane[42] is thermochemically 4.57 ± 0.08 kJ mole^{-1} more stable than the corresponding anti isomer (r-2-t-4-t-5) but the latter is favored in entropy by 4.9 ± 0.3 J °K^{-1} mole^{-1}, in close agreement with the corresponding energy parameters for the equilibrium between cis- and trans-2,5-dimethyl-1,3-oxathiolanes.[32] In the latter case the cis isomer is favored in enthalpy by 4.66 ± 0.32 kJ mole^{-1} and disfavored in entropy by 5.6 ± 1.0 J °K^{-1} mole^{-1}.

TABLE V

CONFORMATIONAL ENTHALPIES OF PSEUDOAXIAL ALKYL
GROUPS IN THE 1,3-OXATHIOLANE RING

	Conformation enthalpy (kJ mole^{-1})		
Alkyl group	CE^a	J_{45} ^{b}	CS^c
2a-Me	4.64	4.90	4.73
2a-Et	4.85	4.98	4.85
2a-i-Pr	5.86	5.69	—
2a-t-Bu	8.37	—	—
5a-Me	—	4.81	
4a-Me	~0		

^{a} From chemical equilibration [R. Keskinen, A. Nikkila, and K. Pihlaja, *Tetrahedron* 28, 3943 (1972); K. Pihlaja, R. Keskinen, and A. Nikkila, Unpublished results dealing with conformational analysis of 2,4-dialkyl-, 2,2,4-trialkyl-, 4,5-dimethyl-, 2,4,5-trimethyl-, and 2,2,4,5-tetramethyl-1,3-oxathiolanes].

^{b} From vicinal coupling constants [R. Keskinen, A. Nikkila, and K. Pihlaja, *Tetrahedron* 28, 3943 (1972)].

^{c} From chemical shifts [D. J. Pasto, F. M. Klein, and T. W. Doyle, *J. Amer. Chem. Soc.* 89, 4368 (1967)].

The 1,3-oxathiolane ring system is thus much less flexible than, for instance, 1,3-dioxolane and 1,3-dithiolane: conformationally, the 1,3-oxathiolane ring is intermediate between a continuously pseudorotating system and the well-defined conformations of six-membered ring systems.

5. *Miscellaneous Five-Membered Systems*

In addition to the five-membered saturated heterocycles discussed above, data are available for 2-alkoxytetrahydrofurans[44,45] and 1,3-oxazolidines.[46,47]

1,3-Oxazolidine and its derivatives can occur as tautomeric mixtures of the open-chain and ring forms:

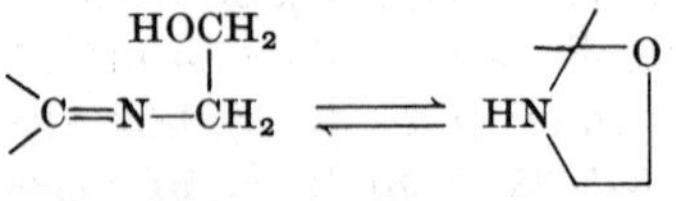

[44] A. Kankaanperä and K. Miikki, *Acta Chem. Scand.* 23, 1471 (1969).
[45] A. Kankaanperä and K. Miikki, *Suom. Kemistilehti B* 41, 42 (1968).
[46] J. V. Paukstelis and R. M. Hammaker, *Tetrahedron Lett.* 3357 (1968); J. V. Paukstelis and L. L. Lambing, *ibid.* 299 (1970).
[47] K. Pihlaja and K. Aaljoki, personal communications (1971).

This equilibrium state is reached very rapidly at different temperatures and is easily followed by NMR[46,47] or IR techniques.[47,48]

Kankaanperä and Miikki[44] studied the cis- trans-2,5-dimethoxytetrahydrofuran equilibrium in various solvent to clarify the role of the anomeric effect in five-membered oxygen heterocycles [Eq. (4)]. Unfortunately, they did not correct free energy differences by 1.72 kJ mole^{-1}, which is required since the trans isomer is a dl pair. With such a correction the cis isomer is thermodynamically more stable in all the solvents used (Table VI) and thus the anomeric effect has no

TABLE VI

EQUILIBRIUM DATA FOR ISOMERIC 2,5-DIMETHOXYTETRAHYDROFURANS TOGETHER WITH SOME COMPARATIVE RESULTS IN DIFFERENT SOLVENTS AT 25°C

Compound	Solvent	$\epsilon_{25}{}^a$	ΔG^0 (cis $\rightleftharpoons$ trans) (kJ mole^{-1})
2,5-Dimethoxytetrahydrofuran	C_6H_6	2.28	0.40[b,c]
	$(C_2H_5)_2O$	4.34	0.42[b,c]
	Neat	—	0.80[b,c]
	CH_3OH	33.62	1.20[b,c]
	$C_6H_5NO_2$	35.74	1.48[b,c]
	H_2O	80.37	2.45[b,c]
2,4-Dimethyl-1,3-dioxolane	$(C_2H_5)_2O$	4.34	1.15[d]
2,6-Diethoxytetrahydropyran	CCl_4	2.24	-3.17[b,e]
	C_2H_5OH	24.30	-2.02[b,e]
	$C_6H_5NO_2$	35.74	-1.97[b,c]
	CH_3CN	37.5	-11.46[b,e]
	H_2O	80.37	-0.97[b,c]
2,5-di-t-Butyl-1,3-dioxane	C_6H_6	2.28	-6.36[f]
	Neat	—	-6.44[f]
	CH_3OH	33.62	-6.78[f]
	$C_6H_5NO_2$	35.74	-6.95[f]
	CH_3CN	37.5	-7.24[f]
	HCOOH	58.5	ab. -8.24[f]

^{a} Dielectric constant at 25°C.

^{b} Corrected in respect to the entropy of mixing of the trans isomers due to the dl forms.

^{c} From A. Kankaanpera and K. Miikki, Acta Chem. Scand. 23, 1471 (1969).

^{d} The entropy favors the cis isomer by 1.1 J K^{-1} mole^{-1} [P. Salomaa, Ann. Univ. Turku, Ser. A No. 46 (1961)].

^{e} E. L. Eliel and C. A. Giza, J. Org. Chem. 33, 3754 (1968).

^{f} E. L. Eliel and D. I. C. Raileanu, Chem. Commun. 290 (1970).

[48] E. D. Bergmann, Chem. Rev. 53, 309 (1953).

significant influence on the equilibrium[44,49] (cf. 2,4-dialkyl-1,3-dioxolanes in Section III,B,2). Nevertheless, the variation of the 2,5-dimethoxytetrahydrofuran equilibrium with the solvent polarity is similar to that of the 2,6-diethoxytetrahydropyrans.[44,50] The magnitude of the change in the equilibrium state also resembles that of the 2-Me-5-t-Bu-1,3-dioxanes in the same solvents.[51]

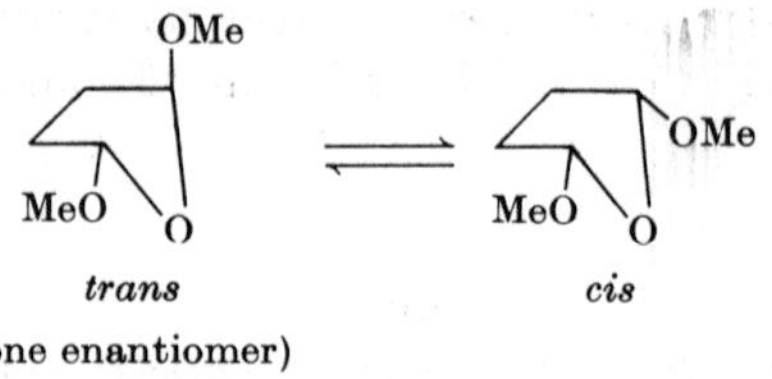

$$
\underset{\substack{trans \\ \text{(one enantiomer)}}}{\text{MeO}\diagdown_O \overset{OMe}{\diagup}} \rightleftharpoons \underset{cis}{\text{MeO}\diagdown_O \diagup OMe} \tag{4}
$$

Later Kankaanperä and Miikki[45] prepared 2,3,5-trimethoxytetrahydrofurans to investigate the possible anomeric effect in these five-membered heterocycles. However, they did not take into account the fact that four stereoisomeric trimethoxy compounds might be involved and hence the equilibration results for a mixture of only three separate isomers without final clarification of their configurations has little significance.[45]

C. Six-Membered Heterocycles

1. Introductory Remarks

Conformational analysis of six-membered saturated heterocycles (26 and 27) has progressed rapidly during the last decade. The precise effects of different heteroatoms on the configurational and conformational properties of six-membered and other heterocycles in general have been needed. Progress in instrumental techniques and the relatively simple synthetic methods needed in the preparation of compounds like 26 and 27 have led to an appreciable increase in their study. Since the techniques of investigation and conclusions drawn from the results of chemical equilibration, are easily applicable to other heterocyclic systems, the following discussion is limited to the above mentioned six-membered systems.

[49] A. Kankaanperä, *Suom. Kemistilehti B* **42**, 208 (1969).
[50] E. L. Eliel and C. A. Giza, *J. Org. Chem.* **33**, 3754 (1968).
[51] E. L. Eliel and D. I. C. Raileanu, *Chem. Commun.* 291 (1970).

2. 1,3-Dioxanes

This ring system is one of the most widely studied six-membered heterocycles. Eliel and his co-workers have extensively investigated alkyl-substituted 1,3-dioxanes by chemical equilibration and NMR spectroscopy.[26,27,50-53a,b] The results have been frequently reviewed[54a-d] and hence we present now only some values for the conformational energies of the axial alkyl groups in different positions of the 1,3-dioxane ring (Table VII).

Pihlaja and his co-workers have also used chemical equilibration and NMR techniques to clarify some characteristic properties of alkyl-substituted 1,3-dioxanes.[55-58a,b] They have also used enthalpies of formation[7,24,25] and appearance potentials measured by electron impact method[28] to study conformations and conformational energies of 1,3-dioxanes. These results are also presented in Table VII.

Riddell and Robinson[59] studied the equilibria between cis- and trans-2-t-Bu-5-alkyl-1,3-dioxanes to determine the conformational energies of the different alkyl groups in position 5.

Very recently Eliel et al.[27,54b,c] pointed out that four-component equilibria may be used to determine conformational energies not available by conventional methods. In this way they were able to estimate a ΔG_{ct} for the chair–twist equilibrium of the 1,3-dioxane ring (see also Section III,D,1). This method also has limitations but nevertheless seems applicable to other systems. Many other reports dealing with the 1,3-dioxane family have also been published.[60-64]

[52] E. L. Eliel and M. K. Kaloustian, *Chem. Commun.* 290 (1970).

[53a] E. L. Eliel and R. M. Enanoza, *J. Amer. Chem. Soc.* **94**, 8072 (1972).

[53b] E. L. Eliel and M. C. Knoeber, *J. Amer. Chem. Soc.* **90**, 3444 (1968).

[54a] E. L. Eliel, *Accounts Chem. Res.* **3**, 1 (1970).

[54b] E. L. Eliel, *Pure Appl. Chem.* **25**, 509 (1971).

[54c] E. L. Eliel, *Int. Congr. Pure Appl. Chem., 23rd* **7**, 219 (1971).

[54d] E. L. Eliel, *Angew. Chem.* **84**, 779 (1972).

[55] K. Pihlaja and P. Äyräs, *Acta Chem. Scand.* **24**, 204, 531 (1970).

[56] K. Pihlaja, G. M. Kellie, and F. G. Riddell, *J. Chem. Soc. Perkin II* 252 (1972).

[57] K. Pihlaja, *Suom. Kemistilehti B* **41**, 229 (1968); **42**, 74 (1969).

[58a] K. Pihlaja and P. Äyräs, *Suom. Kemistilehti B* **43**, 171 (1970).

[58b] K. Pihlaja and A. Tenhosaari, *Suom. Kemistilehti B* **43**, 175 (1970).

[59] F. G. Riddell and M. J. T. Robinson, *Tetrahedron* **23**, 3417 (1967).

[60] M. Anteunis, G. Swaelens, F. Anteunis-De Ketelaere, and P. Dirinck, *Bull. Soc. Chim. Belg.* **80**, 409 (1971).

[61] M. Anteunis and G. Swaelens, *Org. Magn. Res.* **2**, 389 (1970).

[62] E. Coene and M. Anteunis, *Bull. Soc. Chim. Belg.* **79**, 25 (1970).

[63] G. Eccleston, E. Wyn-Jones, and W. J. Orville-Thomas, *J. Chem. Soc. B* 1551 (1971).

[64] For further citations see Eliel,[54a-d] Pihlaja and Äyräs,[55] Pihlaja et al.,[56] Pihlaja,[57] Pihlaja and Äyräs,[58a] and Pihlaja and Tenhosaari.[58b]

TABLE VII

CONFORMATIONAL ENERGIES FOR AXIAL ALKYL GROUPS IN 1,3-DIOXANES,
1,3-DITHIANES, AND 1,3-OXATHIANES

Compound	Alkyl group	Conformational energy (kJ mole^{-1})				Ref.
		2a	4a	5a	6a	
1,3-Dioxane	Me	16.65	12.2	3.35–4.1	12.2	g,h
		17.0^a	11.4^a	—	11.4^a	i
			12.2^a		12.2^a	
		16.3^b	14.0^b	—	14.0^b	i
		19.8^{b,f}				i,j
		15.9^c	14.2^c	—	14.2^c	k
	Et	16.9	—	2.8–3.05	—	h,l
	i-Pr	17.45	—	4.1	—	h,l
	t-Bu	—	—	5.7–6.1	—	h
1,3-Dithiane	Me	(7.4)	(7.1)	(4.35)	(7.1)	m
	Et	(6.4)	—	(3.2)	—	m
	i-Pr	(8.2)	—	(3.3–3.6)	—	m
	t-Bu	(11.4^d)	—	(7.7^d)	—	m
1,3-Oxathiane	Me	13.6	7.45	2.85–3.1	12.3	n
		—	6.5^e	—	10.9^e	o
		—	9.6^{c,f}	—	13.0^{c,f}	o
	Et	13.6	—	—	—	n
	i-Pr	14.85	—	—	—	n
Cyclohexane	Me 7.1				Footnote 5 in m	
	Et 7.3					
	i-Pr 9.0					

^{a} From enthalpies of formation in the liquid state.

^{b} From enthalpies of formation in the gaseous state.

^{c} From appearance potentials of the parentless-methyl ions.

^{d} Twist-boat forms predominate. In fact, all of the conformational energies in 1,3-dithiane should be revised appreciably upwards except that for 4-methyl, which is too high, as will be shown by Pihlaja.[68b,69]

^{e} From ionization potentials.

^{f} For double-butressed molecules.

^{g} For the gaseous state.

^{g} F. W. Nader and E. L. Eliel, J. Amer. Chem. Soc. 92, 3050 (1970).

^{h} E. L. Eliel and M. C. Knoeber, J. Amer. Chem. Soc. 90, 3444 (1968).

^{i} K. Pihlaja and S. Luoma, Acta Chem. Scand. 22, 2401 (1968).

^{j} K. Pihlaja and P. Vainiotalo, private communication, 1972.

^{k} K. Pihlaja and J. Jalonen, Org. Mass Spectrom. 5, 1363 (1971).

^{l} F. W. Nader and E. L. Eliel, J. Amer. Chem. Soc. 92, 3050 (1970).

^{m} E. L. Eliel and R. O. Hutchins, J. Amer. Chem. Soc. 91, 3703 (1969).

^{n} P. Pasanen and K. Pihlaja, Tetrahedron 28, 2617 (1972).

^{o} J. Jalonen, P. Pasanen, and K. Pihlaja, Org. Mass Spectrom. 7, 949 (1973).

3. *1,3-Dithianes*

In comparison with 1,3-dioxanes, relatively little work has been done on the 1,3-dithiane ring system. Eliel and Hutchins[65] studied the conformational preferences of several alkyl-substituted 1,3-dithianes by means of chemical equilibration and NMR. Their results indicated that the 2- and 4-alkyl substituents usually have conformational energies ($-\Delta G^0$) similar to those of the corresponding groups in cyclohexane and thus considerably smaller than those of the same substituents in 1,3-dioxane ring. However, 5-alkyl groups usually have conformational energies relatively close to those of 1,3-dioxane derivatives and thus considerably smaller than the corresponding energies in cyclohexanes (Table VII).

The most interesting result in the reports of Eliel and Hutchins[54a-d,65,66] is the evaluation of the chair–twist enthalpy difference (Section III,D,2). Also Pihlaja and co-workers[67-69] have carried out chemical equilibrations of some epimeric alkyl-substituted 1,3-dithianes (Section III,D,2).

4. *1,3-Oxathianes*

1,3-Oxathianes are very interesting structural intermediates between 1,3-dioxanes and 1,3-dithianes [*33*], the conformational properties of which have been recently clarified with the aid of chemical equilibration.[68,70a,b,71a-d]

$$\langle \text{structure} \rangle \quad \begin{array}{l} \text{1,3-Dioxane, X = Y = O} \\ \text{1,3-Oxathiane, X = O; Y = S} \\ \text{1,3-Dithiane, X = Y = S} \end{array}$$

[*33*]

[65] E. L. Eliel and R. O. Hutchins, *J. Amer. Chem. Soc.* **91**, 3703 (1969).

[66] J. Gelan and M. Anteunis, *Bull. Soc. Chim. Belg.* **78**, 599 (1969).

[67] K. Pihlaja, D. M. Jordan, A. Nikkilä, and H. Nikander, *Adv. Mol. Relaxation Processes* **5**, 227 (1973).

[68a] K. Pihlaja and P. Pasanen, *J. Org. Chem.* **39** (1974) (in press).

[68b] K. Pihlaja, *J. Chem. Soc. Perkin II* (1974) (in press).

[69] K. Pihlaja, A. Nikkilä, and H. Nikander, Unpublished results dealing with the equilibration of epimeric 2,4-diMe-2-alkyl-1,3-dithianes.

[70a] P. Pasanen and K. Pihlaja, *Tetrahedron Lett.* 4515 (1971).

[70b] P. Pasanen and K. Pihlaja, *Tetrahedron* **28**, 2617 (1972).

[71a] N. de Wolf and H. R. Buys, *Tetrahedron Lett.* 551 (1970).

[71b] J. Gelan, G. Swaelens, and M. Anteunis, *Bull. Soc. Chim. Belg.* **79**, 321 (1970).

[71c] J. Gelan and M. Anteunis, *Bull. Soc. Chim. Belg.* **77**, 423 (1968).

[71d] E. Coene and M. Anteunis, *Bull. Soc. Chim. Belg.* **79**, 25 (1970).

The experimentally determined conformational energies of the axial alkyl groups in different positions of the 1,3-oxathiane ring do not deviate appreciably from the values obtainable by comparison with the respective values in 1,3-dioxanes and 1,3-dithianes (Table VII).[70a,b]

The conformational energy of an axial 2-alkyl group does not differ very much from that of an axial 6-methyl group. Hence, for instance, *trans*-2,6-dimethyl-1,3-oxathiane is a mixture of two rapidly inter-converting conformations—the 2a6e and 2e6a forms—whereas the *cis* isomer is a biased molecule. The conformational energy of the axial 2-methyl group was, however, determined directly by equilibrating *r*-2-*trans*-4-*trans*-6- and *r*-2-*cis*-4-*cis*-6-trimethyl-1,3-oxathianes of which the former exists predominantly in the chair form with an axial 2-methyl group.

In the equilibrium between *cis*- and *trans*-2,6-dimethyl-1,3-oxa-thianes [Eq. (5)], let K_{obs} be the observed equilibrium constant, K_{II} the known (see above) equilibrium constant for II with the *cis*

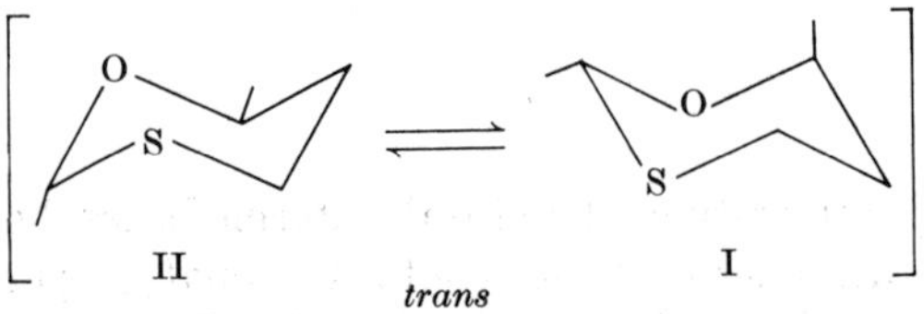

$$\tag{5}$$

II   *trans*   I   *cis*

isomer and K_I the desired equilibrium constant for I with the *cis* isomer.[70a,b,72] Then $K_{obs} = cis/trans$ and consequently, $1/K_{obs} = trans/cis = (I + II)/cis = 1/K_I + 1/K_{II}$. Thus

$$1/K_I = 1/K_{obs} - 1/K_{II} \tag{6}$$

Since the value 13.6 kJ mole^{-1} of the conformational energy of the axial methyl group in the 2a6e conformation (II) was obtained directly,[70b] it is possible, accepting additivity, to calculate the value 12.3 kJ mole^{-1} for the conformational energy of the axial 6-methyl group of I (Table VII). The latter value was justified by further equilibrations.[70b]

The values of the different conformational energies of the axial alkyl groups in different positions of the 1,3-oxathiane ring are summarized in Table VII together with the corresponding values of 1,3-dioxanes and 1,3-dithianes. The chair-twist equilibrium of the 1,3-oxathiane ring system will be discussed in Section III,D,3.

[72] E. L. Eliel and R. S. Ro, *J. Amer. Chem. Soc.* **79**, 5992 (1957).

5. Anomeric Effects in Six-Membered Heterocycles

Eliel and Giza[50] studied the anomeric effect in several 2-alkoxy- and 2-alkylthiotetrahydropyrans [27] and 2-alkoxy-1,3-dioxanes. Kankaanperä and Miikki studied the epimeric 2,6-diethoxytetrahydro-pyrans[44] and Descotes *et al.*[73a,b] and de Hoog *et al.*[74a,b] the 2-alkoxy-tetrahydropyrans by NMR and dipole moments. These results have been repeatedly reviewed[46,54d,75] and are therefore not discussed.

D. Chair–Twist Equilibria in 1,3-Dioxanes, 1,3-Dithianes, and 1,3-Oxathianes

1. 1,3-Dioxanes

The most widely investigated chair–twist equilibrium in six-membered heterocycles is that of the 1,3-dioxane ring.[29] The large enthalpy change and the relatively small entropy change for the chair–twist interconversion in this ring system hinder the use of chemical equilibration.

However, Eliel *et al.*[27,54c] used a four-component equilibrium [Eq. (6)] to solve this problem and obtained 33 kJ mole^{-1} for the $\Delta H_{ct}(l)$ after relevant corrections.[25,27] This agrees well with the other available values—$\Delta H_{ct}(l)$ = 33.2 kJ mole^{-1} from the enthalpies of formation[7] (Section II,C), $\Delta H_{ct}(g)$ = 31.9 kJ mole^{-1} using the group increment method[15] (Section II,C), and $\Delta H_{ct}(g)$ = 35.6 kJ mole^{-1} calculated with the aid of mass spectrometric appearance potentials[28] (*cf.* Table VIII).

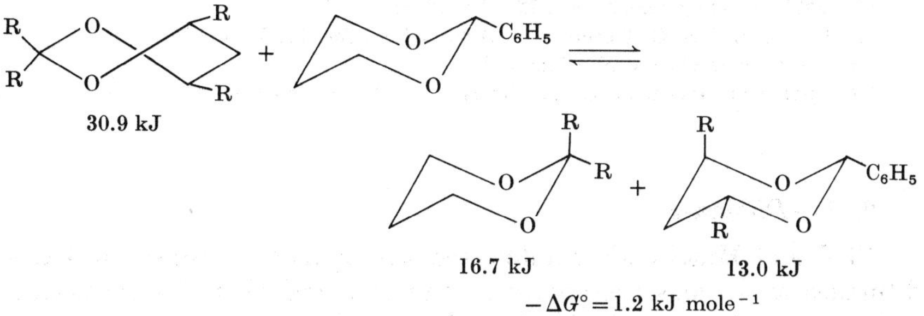

30.9 kJ

16.7 kJ   13.0 kJ

$-\Delta G° = 1.2$ kJ mole^{-1}

R = Me

[73a] M. Gelin, Y. Bahurel, and G. Descotes, *Bull. Soc. Chim. Fr.* 3723 (1970).
[73b] G. Descotes, D. Sinou, and J. C. Martin, *Bull. Soc. Chim. Fr.* 3730 (1970).
[74a] A. J. de Hoog, H. R. Buys, C. Altona, and E. Havinga, *Tetrahedron* 25, 3365 (1969).
[74b] A. J. de Hoog and H. R. Buys, *Tetrahedron Lett.* 4175 (1969).
[75] E. L. Eliel, *Kemisk. Tidskr.* 81, No. 6/7, 22 (1969).

TABLE VIII

CHAIR–TWIST INTERCONVERSION OF 1,3-DIOXANE, 1,3-DITHIANE, AND 1,3-OXATHIANE
IN COMPARISON WITH THAT OF CYCLOHEXANE[a]

Compound	ΔG_{25}^0 (kJ mole^{-1})	ΔH^0 (kJ mole^{-1})	ΔS^0 (J K^{-1} mole^{-1})	Ref.
Cyclohexane[m]	20.5	24.7	14.6	a
1,3-Dioxane[m]	30.9	33.2	7.7[b]	f,g
—		31.9[c]	—	h
—		35.6[c]	—	i
—		37.2	—	j
—	—		16.3	g,k
1,3-Dithiane	7.7[d]	14.3	22.3	l
	14.2[e]	20.2	20.1	m
1,3-Oxathiane	24.6[e]	27.8	10.7	m
—		22.5 $\pm$ 4.0[c]	—	n

[a] See footnote 23 in E. L. Eliel and R. O. Hutchins, J. Amer. Chem. Soc. **91**, 3703 (1969).
[b] Based on the difference between ΔH^0 and ΔG^0 on this line.
[c] For the gaseous state.
[d] For the 2,5-disubstituted series.
[e] For the 2,2,6-trisubstituted compounds. Especially in the case of alkyl-substituted 1,3-dithianes ΔH_{ct} seems to be greatly dependent on the type of substitution.
[f] E. L. Eliel, Pure and Appl. Chem. **25**, 509 (1971).
[g] K. Pihlaja and S. Luoma, Acta Chem. Scand. **22**, 2401 (1968).
[h] K. Pihlaja and P. Vainiotalo, private communication (1972).
[i] K. Pihlaja and J. Jalonen, Org. Mass Spectrom. **5**, 1363 (1971).
[j] G. M. Kellie and F. G. Riddell, "Non-Chair Conformations of Six-Membered Rings," in "Topics in Stereochemistry" (N. L. Allinger and E. L. Eliel, eds.), Vol. 7. Wiley (Interscience), New York, 1973.
[k] K. Pihlaja, Acta Chem. Scand. **22**, 716 (1968).
[l] E. L. Eliel and R. O. Hutchins, J. Amer. Chem. Soc. **91**, 3703 (1969).
[m] For further details see K. Pihlaja.[68a,b]
[n] J. Jalonen, P. Pasanen, and K. Pihlaja, Org. Mass Spectrom. **7**, 949 (1973).

2. 1,3-Dithianes

Eliel and Hutchins[65] equilibrated the epimeric 2,5-di-t-butyl-1,3-dithianes at various temperatures and observed that the cis isomer exists largely in the twist-boat conformation which was confirmed by the temperature dependency of the vicinal coupling constants of cis-2-t-Bu-5-i-Pr- and cis-2,5-di-i-Pr-1,3-dithianes. However, ΔG_{ct} was only about 7.6 kJ mole^{-1} at 25°C, very close to the value 7.4 kJ mole^{-1} derived for the conformational energy of an axial methyl group in position 2. Consequently, it seemed very probable, a priori, that the twist-boat conformation should also make an appreciable contribution to the "biased" equilibria in the determination of the conformational energies of the 4-, 6-, and 2-axial alkyl groups.[65]

To clarify this problem further, Pihlaja and Nikkilä[76] studied the equilibrium between *cis*- and *trans*-2,4-dimethyl-1,3-dithianes [Eq. (7)] at different temperatures. The enthalpy difference of -7.0 kJ mole^{-1} and entropy difference of -5.3 J °K^{-1} mole^{-1} are not far from those expected on the basis of the conformational energies of the 2- and 4-axial methyl groups (7.4 and 7.1 kJ mole^{-1}, respectively[65]; Table VII);

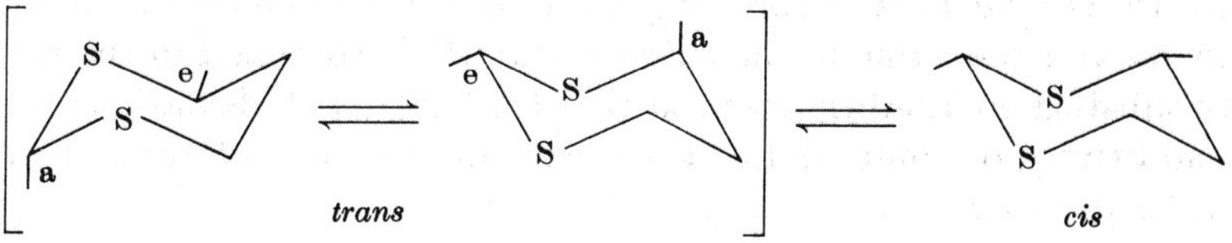

$$\tag{7}$$

but nevertheless indicate that ΔH_{ct} and thus also ΔG_{ct} must be different in value for differently substituted 1,3-dithianes, and accordingly, the parameters determined by Eliel and Hutchins[65] are suitable only for the 2,5-disubstituted twist-boats (Table VIII).[68b,76]

To test the above conclusion the equilibration of *cis*- and *trans*-2,4-dimethyl-2-*t*-butyl-1,3-dithianes was carried out at various temperatures.[68b] The *cis* isomer probably exists in the twist form since its chair conformation includes either an axial *t*-butyl group or a 1,3-transannular nonbonded interaction between axial methyl groups in the 2- and 4-positions. This is supported by the equilibration results which showed that the *trans* isomer is favored in enthalpy by 12.2 kJ mole^{-1}, whereas the entropy change occurs in favor of the *cis* configuration by 20.1 J °K^{-1} mole^{-1} (Table VIII). Taking into account the conformational energy[68b] of the 2-axial methyl group in the *trans* isomer, 20.2 kJ mole^{-1} is obtained for the ΔH_{ct} of the 1,3-dithiane ring: The chair form is thermodynamically (ΔG_{ct}) more stable by 14.2 kJ mole^{-1}. However, the twist form has a significant contribution to those cases where Eliel and Hutchins[65,76] assumed conformationally biased equilibria. Consequently, a full report on the problem of chair–twist equilibria in 1,3-dithianes will be soon published in detail.[68b,76]

3. 1,3-Oxathianes

The chair–twist equilibrium in 1,3-oxathianes probably possess ΔH_{ct} near $0.5(33.5 + 18.6)$, or 26.0, kJ mole^{-1} since in general conformational energies in the 1,3-oxathiane series[29] are near the mean of

[76] K. Pihlaja and A. Nikkilä, Unpublished results.

the respective interaction energies in the correspondingly substituted 1,3-dioxanes and 1,3-dithianes.[70a,b]

Jalonen *et al.*[77] studied the conformational properties of some methyl-substituted 1,3-oxathianes using mass spectrometric ionization and appearance potential measurements as an analytical aid. These measurements led to a value 22.5 ± 4.0 kJ mole^{-1} of the chair–twist enthalpy difference (ΔH_{ct}) in the gaseous state.[68a]

Chemical equilibration of *cis*- and *trans*-2,6-dimethyl-2-*t*-butyl-1,3-oxathianes at various temperatures [Eq. (8)] led, after relevant corrections, to 27.8 kJ mole^{-1} for $\Delta H_{ct}(l)$, in close agreement with the first estimate and the value in the gaseous state.[70,77] Here and in the similarly substituted 1,3-dithianes (Section III,D,2), small deviations from the additivity of conformational energies may occur but these deviations hardly ever exceed 1–2 kJ mole^{-1}.[53,54b]

$$
\underset{\substack{cis \\ (27.8\ \text{kJ})}}{\text{R}\underset{\text{O}}{\overset{\text{S}}{\diagup}}}
\rightleftharpoons
\underset{\substack{trans \\ (14.6\ \text{kJ})}}{\text{R}\underset{\text{O}}{\overset{\text{S}}{\diagup}}}
\qquad (\text{R} = t\text{-Bu}) \qquad (8)
$$

$$-\Delta H^{\circ} = 13.2\ \text{kJ mole}^{-1}$$
$$-\Delta S^{\circ} = 10.7\ \text{J}^{\circ}\text{K}^{-1}\text{mole}^{-1}$$

The energy parameters for the chair–twist equilibria considered in Sections III,D,1–3 are collected in Table VIII with the corresponding data for cyclohexane.

E. Equilibration of *endo–exo* Isomeric Cyclic Vinyl Ethers

Taskinen[78] has recently studied five- and six-membered cyclic vinyl ethers [*28–29*] and determined the equilibrium ratios of the *exo* and *endo*-cyclic isomers and those of the *Z* and *E* isomers at several temperatures. Comparison of the thermodynamic functions for isomerization with those for the corresponding olefins shows that, at equilibrium, the *exo/endo* ratio is more favorable for the *exo* isomers of the five-membered vinyl ethers than for those of the cyclic olefins.[79a–d] In the

[77] J. Jalonen, P. Pasanen, and K. Pihlaja, *Org. Mass Spectrom.* **7**, 949 (1973).
[78] E. Taskinen, PhD Thesis, *Ann. Acad. Sci. Fenn. Ser. A.2* No. 163 (1972).
[79a] A. C. Cope, D. Ambros, E. Ciganek, C. F. Howell, and Z. Jacura, *J. Amer. Chem. Soc.* **82**, 1750 (1960).
[79b] R. B. Turner and R. H. Garner, *J. Amer. Chem. Soc.* **80**, 1424 (1958).
[79c] J. Herling, J. Shabtai and E. Gil-Av, *J. Amer. Chem. Soc.* **87**, 4107 (1965).
[79d] D. B. Bigley and R. W. May, *J. Chem. Soc. B* 1761 (1970).

TABLE IX

THERMODYNAMIC PARAMETERS FOR THE ISOMERIZATION OF SOME CYCLIC VINYL ETHERS IN COMPARISON WITH SIMILAR DATA FOR SOME CYCLIC OLEFINS

Reaction[a]	$-\Delta H°$ (kJ mole^{-1})	$\Delta S°$ (J K^{-1} mole^{-1})
2-Methylenetetrahydrofuran ⇌ 5-methyl-2,3-dihydrofuran	4.7 ± 0.3	8.0 ± 1.0
Z-2-Ethylidenetetrahydrofuran ⇌ 5-ethyl-2,3-dihydrofuran	−2.0 ± 0.3	8.0 ± 1.0
E-2-Ethylidenetetrahydrofuran ⇌ Z-2-ethylidenetetrahydrofuran	3.7 ± 0.3	−1.6 ± 1.3
2-Isopropylidenetetrahydrofuran ⇌ 5-isopropyl-2,3-dihydrofuran	−6.4 ± 0.3	5.5 ± 0.9
Methylenecyclopentane ⇌ 1-methylcyclopentene	16.2	4.2
Ethylidenecyclopentane ⇌ 1-ethylcyclopentene	5.4	6.7
Isopropylidenecyclopentane ⇌ 1-isopropylcyclopentene	−4.3	10.8
2-Methylenetetrahydropyran ⇌ 6-methyl-3,4-dihydro-2H-pyran	13.4 ± 1.4	28.0 ± 6.6
Z-2-Ethylidenetetrahydropyran ⇌ 6-ethyl-3,4-dihydro-2H-pyran	9.1 ± 0.5	16.7 ± 2.1
E-2-Ethylidenetetrahydropyran ⇌ 6-ethyl-3,4-dihydro-2H-pyran	8.2 ± 1.0	34.7 ± 3.3
2-Isopropylidenetetetrahydropyran ⇌ 6-isopropyl-3,4-dihydro-2H-pyran	7.7 ± 0.6	11.3 ± 2.5
Methylenecyclohexane ⇌ 1-methylcyclohexene	10.0	11.7
Ethylidenecyclohexane ⇌ 1-ethylcyclohexene	5.0	1.2
Isopropylidenecyclohexane ⇌ 1-isopropylcyclohexene	3.05	7.0

[a] For further details see E. Taskinen, PhD Thesis, *Ann. Acad. Sci. Fenn. Ser. A. 2* No. 163 (1972), A. C. Cope, D. Ambros, E. Ciganek, C. F. Howell, and Z. Jacura, *J. Amer. Chem. Soc.* **82**, 1750 (1960), R. B. Turner and R. H. Garner, *ibid.* **80**, 1424 (1958), J. Herling, J. Shabtai, and E. Gil-Av, *ibid.* **87**, 4107 (1965), and D. B. Bigley and R. W. May, *J. Chem. Soc. B* 1761 (1970).

six-membered ring compounds the situation is, however, reversed, the *exo/endo* ratio at equilibrium being very unfavorable for the *exo*-cyclic vinyl ethers.

Examples of the energy parameters are given in Table IX with some comparative data for the corresponding cyclic olefins.[78,79a–d]

IV. RING STRAIN IN HETEROCYCLIC COMPOUNDS

A. Introduction

It is well known that the energy content of a methylene group in a cycloalkane depends considerably on ring size. The lowest energy content per methylene group in cycloalkanes up to cyclohexadecane is found in cyclohexane, whose heat of combustion per CH_2 group is equal to that of a long straight-chain alkane. Complete staggering of methylene groups is possible in straight-chain paraffins, and it is customary to regard these molecules as "strain free." Thus cyclohexane might also be considered strain free, whereas all the other cycloalkanes up to $C_{16}H_{32}$ have positive strain energies.

The strain in unsubstituted saturated compounds derives from several sources. The strain energy of three- and four-membered rings is mainly due to deformation of the normal bond angles of the atoms forming the ring. Hence this kind of strain is called *angular strain*. In medium-sized rings *interaction strain* between neighboring methylene groups is mostly responsible for the strain observed, while interaction between hydrogen atoms on opposite sides of a large-ring compound (*transannular strain*) may contribute significantly to the total strain. In cyclic compounds containing unsaturation there is an additional possibility of *torsional strain* or twisting of the olefinic link to permit ring formation (note: interaction strain between nonbonded atoms has also been termed torsional strain by several authors).

If the ring contains substituents, the total strain can arise from the ring itself, from interactions of the substituents with the ring, and from mutual interactions between two or more substituents. Thus the total strain may be much higher than the strain of the unsubstituted ring, and hence a distinction between these two quantities should be made in discussions of ring strain energy.

Ring strain is usually evaluated by comparing experimental heats of formation or atomization with those calculated by means of some bond energy or group increment scheme. Because of the many different bond energy or group increment schemes available at present, considerable differences in strain energy may be encountered when strain energies from different sources are compared. Hence it would be desirable to have only one bond energy or group increment scheme in use. All the recalculated strain energy values reported in this section are based on the equivalent bond energy or group increment schemes given by Cox and Pilcher,[1] or, in a few cases (see Tables X and XI), using the Allen-type scheme described in Section II,B. Following the terminology of Cox and Pilcher,[1] ring strain energies calculated by means of these schemes are referred to as conventional ring strain energies (C.R.S.E.s).

Another method for estimating ring strain energies is to compare the heat of hydrogenation of the compound concerned with that of a reference compound. The latter should be selected with care to avoid neglecting strain that may be present in the reference compound or its hydrogenation product. For example, the strain in cyclopentene might be evaluated by comparing its heat of hydrogenation to cyclopentane with that of trans-2-butene to butane. The last two molecules are generally regarded as strain free and thus comparison of the two heats of hydrogenation should give the difference in strain between cyclopentene and cyclopentane. If the strain energy of cyclopentane is known, the strain in cyclopentene is easily obtained.

For a more detailed treatment of the concept of ring strain, readers are referred to an article of Nelander and Sunner[80] where the effects of pressure–volume energy, nuclear motions, and reference system are discussed.

Introduction of a heteroatom into a carbocyclic ring may affect the ring strain in several ways. Replacement of a carbon atom by a heteroatom whose covalent radius is shorter than that of carbon obviously increases angular strain in small-ring compounds. On the other hand, nonbonded interactions might be expected to decrease as two methylene–methylene interactions are removed. The amount of this kind of strain release naturally depends on the extent of eclipsing of the methylene groups in the parent carbocyclic compound. In most cases the replacement of a methylene group by an oxygen or a nitrogen atom causes only a small change in ring strain, but in sulfur-containing heterocycles strain release may be appreciable owing to the relatively large covalent radius of sulfur.

B. Ring Strain in Saturated Heterocycles with One Heteroatom

Conventional ring strain energies of some monocyclic compounds with one heteroatom have been collected in Table X. For the sake of comparison, ring strain energies of the corresponding carbocyclic compounds

[80] B. Nelander and S. Sunner, J. Chem. Phys. 44, 2476 (1966).

TABLE X

CONVENTIONAL RING STRAIN ENERGIES (kJ mole⁻¹) OF SOME MONOCYCLIC RING COMPOUNDS WITH ONE OR NO HETEROATOM

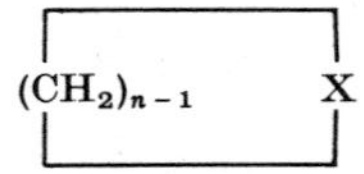

$$\overset{X}{\underset{(CH_2)_{n-1}}{\boxed{}}}$$

X	Enthalpy quantity	n (number of atoms in the ring)									
		3		**4**		**5**		**6**		**7**	
		G^a	A^b	G	A	G	A	G	A	G	A
CH_2	ΔH_f° (est.)	−61.8	−62.7	−82.3	−83.6	−102.9	−104.5	−123.5	−125.4	−144.1	−146.3
	ΔH_f° (exptl.)	+53.3	+53.3	+28.4	+28.4	−77.2	−77.2	−123.4	−123.4	−118.0	−118.0
	C.R.S.E.	115.1	116.0	110.7	112.0	25.7	27.3	0.1	2.0	26.1	28.3
O	ΔH_f° (est.)	−166.5	−165.9	−187.1	−186.8	−207.7	−207.7	−228.3	−228.6	—	—
	ΔH_f° (exptl.)	−52.6	−52.6	−80.5	−80.5	−184.2	−184.2	−223.4	−223.4	—	—
	C.R.S.E.	113.9	113.3	106.6	106.3	23.5	23.5	4.9	5.2	—	—
NH	ΔH_f° (est.)	+13.0	+14.0	—	—	−28.2	−27.8	−48.8	−48.7	—	—
	ΔH_f° (exptl.)	+126.5	+126.5	—	—	−3.4	−3.4	−49.2	−49.2	—	—
	C.R.S.E.	113.5	112.5	—	—	24.8	24.4	−0.4	−0.5	—	—
S	ΔH_f° (est.)	−1.0	+0.5	−21.6	−20.4	−42.1	−41.3	−62.7	−62.2	−83.3	−83.1
	ΔH_f° (exptl.)	+82.4	+82.4	+61.0	+61.0	−33.8	−33.8	−63.3	−63.3	−65.5	−65.5
	C.R.S.E.	83.4	81.9	82.6	81.4	8.3	7.5	−0.6	−1.1	17.8	17.6

^{a} Group method.
^{b} Allen method.

228

are also included in the table. The experimental standard heats of formation of the gaseous cyclanes are the "selected values" of the heats of formation given by Cox and Pilcher.[1] Calculated heats of formation were evaluated by the Group method of the above authors and, for comparison, by the Allen scheme described in Section II,B.

The data of Table X show that ring strain energies of three- and four-membered rings are rather similar (about 110 kJ mole^{-1}) in those cases where the group X in the ring is CH_2, O, or NH (data for the four-membered nitrogen heterocycle are not available). This is not surprising in view of the almost equal single-bond covalent radii of carbon, oxygen, and nitrogen atoms (77, 74, and 74 pm,[81] respectively). The larger sulfur atom (covalent radius 104 pm[81]) causes a significant release of angular strain in these small-size rings where the ring strain energy amounts only to about 83 kJ mole^{-1}. The same trend is also observable in other ring sizes. It is noteworthy that tetrahydropyran is slightly strained (4.9 kJ mole^{-1}) while all the other six-membered ring compounds in Table X are essentially unstrained.

Alkyl groups are reported to stabilize small-ring compounds. Thus, it has been maintained[82a-c] that the methyl group of 2-methylthiacyclopropane [34] decreases the strain energy of the thiacyclopropane ring with about 16 kJ mole^{-1}, and the introduction of a second methyl group on the same carbon leads to a further decrease of 2–3 kJ mole^{-1}.

This statement disagrees with the C.R.S.E. of 34, 76.4 kJ mole^{-1}, calculated from the experimental[1] and estimated heats of formation (+46.1 and −30.3 kJ mole^{-1}, respectively). The stabilizing effect of a single methyl group on the thiacyclopropane ring is seen to be only about 7 kJ mole^{-1}. The difference 74.8 kJ mole^{-1} between the experimental[1] and estimated heats of formation (+11.6 and −63.2 kJ mole^{-1}, respectively) of 2,2-dimethylthiacyclopropane [35] shows that the stabilizing effect of a second methyl group on the same carbon atom in a thiacyclopropane ring amounts to 1–2 kJ mole^{-1}. A similar comparison between the experimental[1] and estimated heats of formation

[81] G. W. Wheland, "Resonance in Organic Chemistry," p. 170. Wiley, New York, 1955.
[82a] S. Sunner, Studies in Combustion Calorimetry Applied to Organo-Sulfur Compounds (Diss.), Lund University, 1949.
[82b] S. Sunner, Acta Chem. Scand. 17, 728 (1963).
[82c] B. Ringner, S. Sunner, and H. Watanabe, Acta Chem. Scand. 25, 141 (1971).

($+3.9$ and -59.7 kJ mole^{-1}, respectively) of gaseous *trans*-2,3-dimethylthiacyclopropane [*36*] leads to a strain energy of 63.6 kJ mole^{-1}, which shows a stabilization of about 20 kJ mole^{-1} in this compound relative to unsubstituted thiacyclopropane. Mutual destabilizing interaction between the three methyl groups of 2,2,3-trimethylthiacyclopropane [*37*] causes the experimental heat of formation, -21.2 kJ mole^{-1},[1] to be some 71 kJ mole^{-1} more positive than the estimated heat of formation, calculated by means of the Group scheme (with no steric corrections).

Strain energies in *34–36* can also be estimated using the principle of structural similarity. The difference of 30.9 kJ mole^{-1} between the experimental[1] heats of formation of 2-thiabutane and 3-methyl-2-thiabutane (-59.3 and -90.2 kJ mole^{-1}, respectively) is 5.4 kJ lower than the difference, 36.3 kJ mole^{-1}, between the heats of formation of thiacyclopropane and [*34*]. This shows a stabilization of 5.4 kJ mole^{-1} in the latter compound. The heat of formation of 3-thiapentane is 59.5 kJ mole^{-1} higher than that of 2,4-dimethyl-3-thiapentane,[1] while the corresponding difference between thiacyclopropane and [*36*] is 78.5 kJ mole^{-1}. Hence the stabilizing effect of the two methyl groups in [*36*] is about 19 kJ mole^{-1}. A similar reasoning using the heats of formation[1] of 2-thiabutane and 3,3-dimethyl-2-thiabutane, on one hand, and the heats of formation of thiacyclopropane and [*35*], on the other, leads to a stabilization of about 9.1 kJ mole^{-1} in [*35*].

The stabilizing effect of a methyl group on the corresponding oxygen heterocycle, ethylene oxide, is obtained by comparing the C.R.S.E., 110.1 kJ mole^{-1}, of propylene oxide [*38*] [from ΔH_f^0(exptl.) $= -94.7$[1] and ΔH_f^0(est.) $= -204.8$ kJ mole^{-1}], with that of the parent ethylene oxide ($= 113.9$ kJ mole^{-1}, Table X). The stabilization is about 4 kJ mole^{-1}. The difference, 35.6 kJ mole^{-1}, between the heats of formation

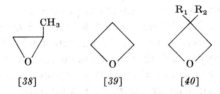

[*38*]   [*39*]   [*40*]

of gaseous methyl ethyl ether and methyl isopropyl ether (-216.4 and -252.0 kJ mole^{-1},[1] respectively) as compared with the difference (42.0 kJ mole^{-1}) between the heats of formation of ethylene oxide and *38* reveals a relative stabilization of 6.4 kJ mole^{-1} in the latter compound.

Ringner *et al.*[82c] have shown that the strain in *40* ($R_1 = R_2 = CH_3$; $R_1 = R_2 = CH_2Cl$; $R_1 = Et$, $R_2 = CH_2Cl$) is some 13 kJ mole^{-1} lower than the strain (106.6 kJ mole^{-1}) in the parent oxetane [*39*]. Their calculation was based on the principle of structural similarity. If the strain energies of the three compounds [*40*] are estimated by means of the Group scheme, it is seen that only when $R_1 = R_2 = CH_3$ do these two methods give equivalent values for the strain energy, if no steric corrections are made. For the other two oxetanes, strain energies of about 103 kJ mole^{-1} are obtained but, including four alkane 1,4-*gauche* interactions (2.5 kJ mole^{-1} each) in the estimated heats of formation of the latter compounds, the strain energies in these molecules are seen to be about 14 kJ mole^{-1} lower than the strain in [*39*].

The experimental heat of formation, -63.3 kJ mole^{-1},[1] of gaseous 2-methylthiacyclopentane [*10*] is 8.2 kJ mole^{-1} higher than expected using the Group scheme parameters. This value of the strain energy in [*10*] is equal to the strain in the parent thiacyclopentane. Similar strain energies in these compounds are also confirmed by the heats of formation of thiacyclopentane and *10* which differ by practically the same amount (29.5 kJ mole^{-1}) as do those of their open-chain analogs, 2-thiabutane and 3-methyl-2-thiabutane (30.9 kJ mole^{-1}). The stabilizing effect of alkyl groups on ring compounds is seen to be limited to three- and four-membered rings.

Bedford *et al.*[83] have determined the heats of formation of liquid 7-oxabicyclo[2.2.1]heptane [*41*] and its 2-*exo*- and 2-*endo*-methyl derivatives. The strain in [*41*] was estimated to be about 27 kJ mole^{-1}, some 50 kJ less than the strain in the corresponding hydrocarbon, bicyclo[2.2.1]heptane. The strain in [*41*] can also be estimated by

[*41*]

comparing its experimental heat of formation in the gas phase with the heat of formation obtained from the Group scheme. From the equation of Wadsö,[84] the heat of vaporization of [*41*] (b.p. 119–119.5°C) at 25°C can be estimated to be about 41 kJ mole^{-1}. Thus the standard heat of formation of gaseous [*41*] is about -183 kJ mole^{-1}, which value is about 14 kJ mole^{-1} higher than the heat of formation (-196.6 kJ

[83] A. F. Bedford, A. E. Beezer, C. T. Mortimer, and H. D. Springall, *J. Chem. Soc.* 3823 (1963).
[84] I. Wadsö, *Acta Chem. Scand.* **20**, 544 (1966).

mole^{-1}) estimated using the Group scheme. Both of the estimated strain energies of [41] are surprisingly much lower than the strain energy in either bicyclo[2.2.1]heptane[83] or 2 moles of tetrahydrofuran, 77 kJ mole^{-1} and 47 kJ mole^{-1}, respectively.

C. Ring Strain in Saturated Heterocycles with Two or More Heteroatoms

Thermochemical data on saturated heterocycles with two or more heteroatoms is limited and deals mainly with oxygen heterocycles. Ring strain energies, defined as the difference between experimental and estimated heats of formation of the gaseous compounds in their standard states at 298°K, are listed in Table XI.

The stabilizing effect of a 1,3-dioxa-grouping (—O—C—O—) rela-

TABLE XI

CONVENTIONAL RING STRAIN ENERGIES (kJ mole^{-1}) OF SOME SATURATED HETEROCYCLIC COMPOUNDS WITH TWO OR MORE HETEROATOMS

| Compound | ΔH_f^0(est.) | | ΔH_f^0(exptl.) | C.R.S.E. | |
	G[a]	A[b]		G[a]	A[b]
1,3-Dioxolane [18]	−321.1[c]	−321.7[c]	−301.9[d]	19.2	19.8
1,3-Dioxane [17]	−351.3	−350.9	−349.8[e]	1.5	1.1
1,4-Dioxane [22]	−314.0[f]	−316.1	−315.9	−1.9	0.2
1,3-Dioxacycloheptane [42]	−371.8	−371.8	−346.4	25.4	25.4
1,3-Dioxacyclo-octane [43]	−392.4	−392.6	−336.8	55.6	55.8
Trioxane [15]	−492.4	−492.0	−464.5[g]	27.9	27.5
Paraldehyde [23]	−617.8	−623.9	−645.2	−26.7	−21.3
5,5'-Spirobis-1,3-dioxane [44]	−661.6	−659.8	−629.3	32.3	30.5
Tetroxane [16]	−656.6	−656.0	−620.2[h]	36.4	35.8

[a] Group method.

[b] Allen method.

[c] Corrected with the interaction energy of a 1,4-dioxa-grouping.

[d] A mean of the ΔH_f^0 values given in J. D. Cox and G. Pilcher, "Thermochemistry of Organic and Organometallic Compounds," ref. 59/20. Academic Press, New York, 1970; K. Pihlaja and J. Heikkilä, Acta. Chem. Scand. 23, 1053 (1969).

[e] A mean of the ΔH_f^0 values given in J. D. Cox and G. Pilcher, "Thermochemistry of Organic and Organometallic Compounds." Academic Press, New York, 1970, and K. Pihlaja and S. Luoma, Acta Chem. Scand. 22, 2401 (1968).

[f] Corrected with twice the interaction energy of a 1,4-dioxa-grouping.

[g] A mean of the ΔH_f^0 values given in M. Månsson, E. Morawetz, Y. Nakase, and S. Sunner, Acta Chem. Scand. 23, 56 (1969), M. Delepine and M. Badoche, C. R. Acad. Sci. Paris 214, 777 (1942), and W. K. Busfield and D. Merigold, J. Chem. Soc. A 2975 (1969); Makromol. Chem. 138, 65 (1970).

[h] M. Månsson, E. Morawetz, Y. Nakase, and S. Sunner, Acta Chem. Scand. 23, 56 (1969).

tive to aliphatic monoethers has been long known.[85,86] Månsson[16] has reported a value of 16.9 $\pm$ 0.2 kJ mole^{-1} for this interaction energy. The presence of appreciable amounts of apparent strain in 1,4-dioxane [22] in relation to 1,3-dioxane [17] or acyclic monoethers is also a well-known fact.[87] It has been recently shown by Månsson[16] that the existence of a 1,4-dioxa-grouping even in straight-chain aliphatic ethers causes a "strain" of about 9.7 kJ mole^{-1}. The interaction energy of a 1,5-dioxa-grouping has also been studied by the same author,[88] and it was found that this energy term was zero within the limits of experimental error. As the destabilizing interaction of the 1,4-dioxa-grouping is present even in acyclic ethers containing this structural unit, it was considered justified to exclude this energy contribution from the strain energy values of 1,3-dioxolane [18] and 1,4-dioxane. As the latter compound has two interactions of this kind, twice the destabilizing interaction energy of this grouping has been added to the estimated heat of formation of [22]. The value of this energy contribution was evaluated from the experimental[16] and estimated heats of formation of 3,6-dioxaoctane, and was fixed at 9.5 kJ mole^{-1} (using the Group parameters) or 7.9 kJ mole^{-1} (using the Allen parameters, Section II,B) per 1,4-dioxa-grouping. After this correction, ring strain in 22 is seen to be negligible. The same holds for the isomeric 17.

Both trioxane [15] and paraldehyde [23] show unexpected features (cf. Section II,C). The strain in trioxane is about 27 kJ mole^{-1}, while the latter compound is stabilized by a comparable amount. The strong stabilization (54 kJ mole^{-1}) in 23 relative to 15 is surprising, and indicates a stabilization energy of 18 kJ mole^{-1} per methyl group in 23.

On the other hand, the heat of formation of tetroxane [16] given by Månsson et al.[6] leads to a quite normal strain energy value, 36 kJ mole^{-1}, for an eight-membered ring (C.R.S.E.s in cyclo-octane[1] and 1,3-dioxa-cyclo-octane [43] about 40 and 56 kJ mole^{-1}, respectively). The strain 25.4 kJ mole^{-1} in 1,3-dioxacycloheptane [42] is almost equal to the

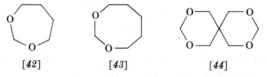

[42]   [43]   [44]

[85] M. S. Kharasch, J. Res. Nat. Bur. Std. 2, 359 (1929).

[86] G. W. Wheland, "Resonance in Organic Chemistry," pp. 89, 95. Wiley, New York, 1955.

[87] A. Snelson and H. A. Skinner, Trans. Faraday Soc. 57, 2125 (1961).

[88] M. Månsson, Acta Chem. Scand. 26, 1707 (1972).

strain (26.1 kJ mole^{-1}, Table X) in the corresponding carbocycle, cycloheptane. Ring strain energy in 1,3-dioxolane (19.2 kJ mole^{-1}) is well related to the strain energy in tetrahydrofuran or cyclopentane (23.5 and 25.8 kJ mole^{-1}, respectively, Table X).

Greatly varying values for the strain in the 1,2-dithiolane ring [45] have been proposed by different authors, as summarized by Bergson and Schotte.[89] From thermochemical measurements, Sunner[90] has estimated the strain in this ring to be about 17–21 kJ mole^{-1}, but Bergson and Schotte[89] suggest a minimum value of 67 kJ mole^{-1} on the basis of conformational analysis of the 1,2-dithiolane ring. If this ring really is

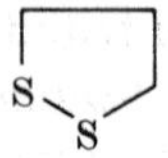

[45] 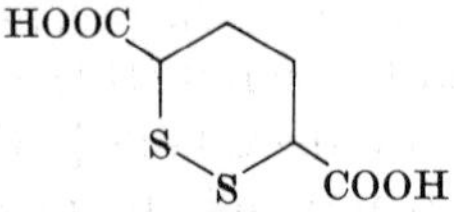   [46]

highly strained, it must be caused by a strong mutual repulsion between the lone-pair orbitals of the neighboring sulfur atoms (see Section IV,D). The strain energy of 1,2-dithiane-3,6-dicarboxylic acid [46] has been estimated[89] to exceed 21 kJ mole^{-1}. The estimation was based on conformational analysis.

Månsson et al.[91] have determined the enthalpy of formation of hexamethylenetetramine [47]. An attempt to estimate the strain in this molecule failed because of the nonexistence of reliable thermochemical data to establish a value for the N—C—N interaction parameter (Γ_{NCN}). Using an Allen type scheme with the Sunner-Nelander[80] ring closure corrections, the strain in 47 was evaluated to be ($6\Gamma_{NCN} - 184$) kJ mole^{-1}.

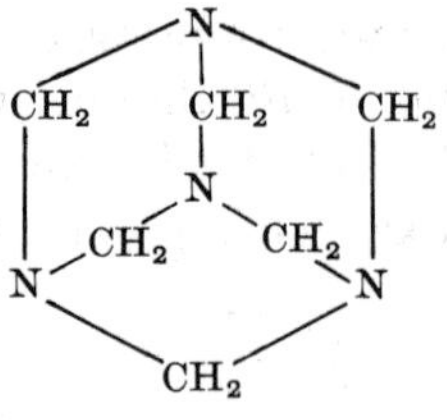

[47]

Thermochemistry of some borolanes, dithioborolanes, borinanes, and dithioborinanes has been studied by Finch et al.[92a,b] The following

[89] G. Bergson and L. Schotte, Ark. Kemi **13**, 43 (1958).

[90] S. Sunner, Nature (London) **176**, 217 (1955).

[91] M. Månsson, N. Rapport, and E. F. Westrum, Jr., J. Amer. Chem. Soc. **92**, 7296 (1970).

[92a] A. Finch and P. J. Gardner, J. Chem. Soc. 2985 (1964).

[92b] A. Finch, P. J. Gardner, and E. J. Pearn, Trans. Faraday Soc. **62**, 1072 (1966).

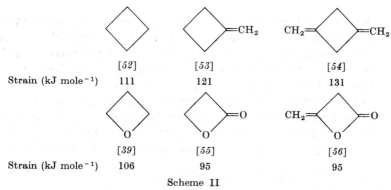

[48]   [49]   [50]   [51]

rough estimates for the strain energies of the compounds *48–51* were given (in kJ mole^{-1}): *48*, 40–80; *49*, 13 ± 18; *50*, 38 ± 22; *51*, 12 ± 18.

D. Ring Strain in Unsaturated Heterocyclic Compounds

Separation of the relative contributions of ring strain and resonance energy on the heats of formation of unsaturated heterocyclic compounds showing strong interaction between the heteroatom(s) and double bond(s) affords insurmountable difficulties. Even in molecules where this interaction energy is relatively low, such as vinyl ethers, quantitative evaluation of the contributions of the two effects on the heat of formation may be impossible. Clearly, most reliable values for the ring strain energy are obtained in those cases where there is no possibility for resonance.

Börjesson *et al.*[93] have estimated the strain energy of β-propiolactone (*55*) to be about 106 kJ mole^{-1}, which value was later corrected by Månsson *et al.*[94] to 94 kJ mole^{-1}. The latter authors estimated the strain in diketene [*56*] to be equal to the strain in [*55*]. The strain energy value assigned to [*55*] by Börjesson *et al.* was based on structural similarities with alkyl alkanoates. By means of the Group scheme parameters and the experimental heat of formation of [*55*], the strain in [*55*] can be calculated to be 95 kJ mole^{-1}. Now it is interesting to compare these strain energy values, together with the C.R.S.E. in oxetane [*39*],

	[*52*]	[*53*]	[*54*]
Strain (kJ mole^{-1})	111	121	131

	[*39*]	[*55*]	[*56*]
Strain (kJ mole^{-1})	106	95	95

Scheme II

[93] B. Börjesson, Y. Nakase, and S. Sunner, *Acta Chem. Scand.* **20**, 803 (1966).
[94] M. Månsson, Y. Nakase, and S. Sunner, *Acta Chem. Scand.* **22**, 171 (1968).

with the strain in cyclobutane [52], methylenecyclobutane [53], and 1,3-dimethylenecyclobutane [54]. Strain energy values for the two latter compounds were evaluated by using the heat of formation data given by Benson *et al.*[95] The results are shown in Scheme II. Replacement of sp^3-hybridized carbon atoms with sp^2-hybridized carbon atoms is seen to increase the strain in the four-membered carbocyclic ring, while the strain of the oxetane ring is reduced on introduction of an sp^2-hybridized carbon atom into the ring. Two sp^2-hybridized carbon atoms in the four-membered oxygen heterocycle seem to have no effect on ring strain, as compared with the oxetane ring with one sp^2-hybridized atom. As the resonance effect of the carbonyl group with the oxygen atom (as in 55) has been taken into account in the enumeration of the Group parameters, the difference between the relative strain energies of 55 and 53 (as compared with the strain energy in oxetane and cyclobutane) cannot be considered as arising from the resonance in 55. However, in view of the about 10 kJ higher strain in 54 relative to that in 53, the equal apparent strain energies of 55 and 56 might be interpreted as an additional resonance stabilization in diketene. The magnitude of this additional stabilization is very reasonable in the light of the stabilization energy, about 15 kJ mole^{-1},[96] present in acyclic vinyl ethers.

Strain energies of the five-membered sulfur heterocycles, 2-thiolene [57] and 3-thiolene [58], have been estimated by Davies and Sunner.[97]

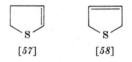

[57]   [58]

They found that the calculated heat of hydrogenation of 58 to thiacyclopentane was almost the same as the heat of hydrogenation of *cis*-2-butene to butane and they concluded that the strain energies of the two ring compounds are equal. This is an erroneous conclusion, because the similar heats of hydrogenation merely show that the difference in strain between 58 and thiacyclopentane is equal to that between *cis*-2-butene and butane. It is a well-known fact that the strain in *cis*-2-butene is about 4.2 kJ mole^{-1}, and thus it may be stated that the strain in 58 exceeds the strain in thiacyclopentane by 4.2 kJ mole^{-1} (if the

[95] S. W. Benson, F. R. Cruickshank, D. M. Golden, G. R. Haugen, H. E. O'Neal, A. S. Rodgers, R. Shaw, and R. Walsh, *Chem. Rev.* **69**, 279 (1969).

[96] G. W. Wheland, "Resonance in Organic Chemistry," p. 85. Wiley, New York, 1955.

[97] J. V. Davies and S. Sunner, *Acta Chem. Scand.* **16**, 1870 (1962).

two heats of hydrogenation are taken to be exactly equal). A more accurate value of the strain energy in *58* is obtained as follows.

The heat of formation of *58* is 121.0 kJ mole^{-1} higher (more positive) than that of thiacyclopentane, while the difference between the heats of formation of *trans*-2-butene ("strain free") and n-butane is 114.5 kJ mole^{-1}.[1] Thus the strain in *58* is $121.0 - 114.5 + 8.4 = 15$ kJ mole^{-1} (8.4 is the strain in thiacyclopentane in kJ mole^{-1}). The Laidler parameters for the estimation of $\Delta H_f{}^0$ given in Cox and Pilcher[1] lead to a strain energy value of 15.8 kJ mole^{-1} for *58*. It is noteworthy that the strain in the five-membered sulfur heterocycles is higher in the unsaturated compound while cyclopentene is about 2 kJ mole^{-1} less strained than cyclopentane.[1]

The strain energy in *57* was estimated by Davies and Sunner[97] to be 12–16 kJ mole^{-1} higher than the strain in *58*. It seems that this estimate needs revision. According to the Laidler parameters given by Cox and Pilcher,[1] the contributions of C_d—C and C—C* bonds to the heat of formation are -19.33 and -0.13 kJ mole^{-1}, respectively. Assuming that the same difference (19.2 kJ mole^{-1}) exists between the contributions of C_d—O and C—O bonds, the experimental[1] and estimated heats of formation of ethyl vinyl ether are found to be equal. Hence it is likely that no large error is introduced if the same difference is assumed between the contributions of C_d—S and C—S bonds. From the experimental and estimated heats of formation (91.1 and 73.6 kJ mole^{-1}, respectively), a strain energy of 18 kJ mole^{-1} may thus be derived for *57*. Hence both isomers are equally strained.

The difference in strain energy between 1,2,5,6-tetrahydropyridine [*59*] and piperidine [*2*] has been estimated by Bedford *et al.*[98] to be the

[*59*]

same as that between cyclohexene and cyclohexane. A lack of experimental heat of vaporization data for *59* forced the authors to base their deductions on liquid-phase heat of formation data. However, it is likely that the heat of vaporization of *59* is very similar to those of pyridine or piperidine, given by Cox and Pilcher.[1] Assuming a value of 39.8 kJ mole^{-1} to represent the heat of vaporization of liquid *59*, the

* C_d means a trigonal carbon atom.

[98] A. F. Bedford, A. E. Beezer, and C. T. Mortimer, *J. Chem. Soc.* 2039 (1963).

heat of formation of gaseous *59* is calculated to be about 73.3 kJ mole^{-1} at 298°K. Then a comparison of the heats of hydrogenation of *59* to piperidine and *trans*-2-butene to n-butane (-122.5 and -114.5 kJ mole^{-1}, respectively) shows that the strain in *59* is about 8.0 kJ mole^{-1} higher than the strain in *2*. Thus a strain energy of about 7.6 kJ mole^{-1} is obtained for *59*. A calculation using the Group parameters gives a value of 8.4 kJ mole^{-1} for this strain energy. On average, ring strain in *59* is about 8 kJ mole^{-1}.

Cass *et al.*[99] have studied the extent of resonance in *60–63*. Their results may also be interpreted in terms of the strain energy in these

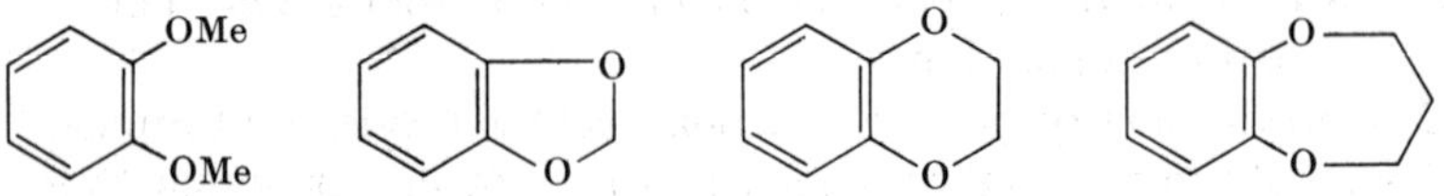

[60]   [61]   [62]   [63]

compounds. Using the Laidler parameters by Cox and Pilcher,[1] the estimated heat of formation of *60* is found to be about 3.8 kJ lower than the experimental value. Taking this energy difference as an *ortho* correction in *60* and applying the same correction for *61–63*, the experimental heats of formation of *61, 62*, and *63* are found to be about 61, 0, and 41 kJ mole^{-1}, respectively, more positive than the heats of formation calculated by means of the Laidler parameters. The heterocyclic ring in *62* is seen to be unstrained. As the strain energies in the five- and seven-membered carbo- and heterocycles discussed earlier in this article were found to be about 20–25 kJ mole^{-1}, it is evident that the large differences between the experimental and estimated heats of formation of *61* and *63* cannot be considered arising entirely from strain but, as suggested by Cass *et al.*,[99] from a weakening of the resonance between the benzene ring and heteroatoms owing to the unfavorable bond angles C_b—O—C in *61* and *63*.*

Heats of combustion for *64–69* have been measured by Geiseler

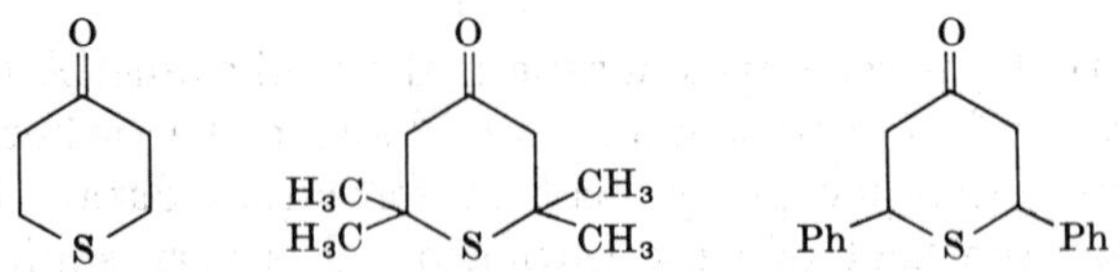

[64]   [65]   [66]

* C_b refers to a carbon atom of the benzene ring.

[99] R. C. Cass, S. E. Fletcher, C. T. Mortimer, P. G. Quincey, and H. D. Springall, *J. Chem. Soc.* 2595 (1958).

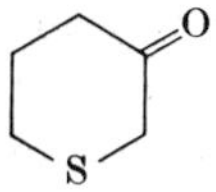

[67] 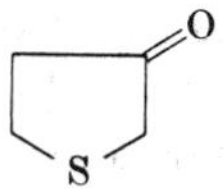   [68] 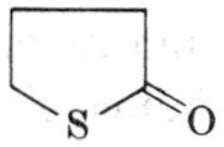   [69]

and Sawistowsky.[100] It was found that *64–68* were strained, while *69* showed a strong stabilization, about 64 kJ mole^{-1}, obviously due to an enhanced resonance between the carbonyl group and sulfur atom. On the other hand, destabilizing interaction between these groups in *64, 67*, and *68* resulted in strain energies of about 14, 36, and 14 kJ mole^{-1}, respectively, in these compounds.

Geiseler and Rauh[101] have evaluated energy differences between the experimental and estimated heats of formation of *70–72*. All these

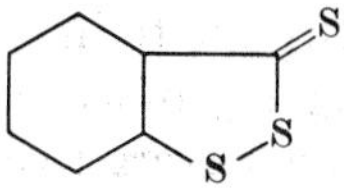

[70] 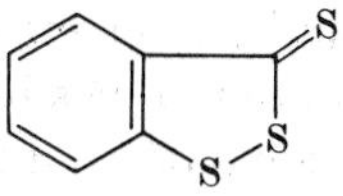   [71] 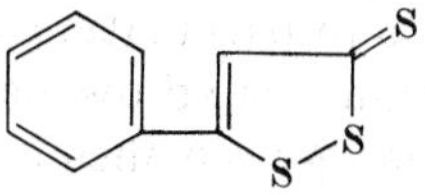   [72]

compounds exhibited an energy content about 70 kJ mole^{-1} higher than predicted using Cox's bond energy scheme.[102] The high strain was interpreted to be due to a strong repulsion between the lone-pair orbitals of neighboring sulfur atoms, the orbitals being forced to lie in almost the same plane by the five-membered ring. The results seem to confirm the high strain energy in the 1,2-dithiolane ring (Section IV,C).

V. RESONANCE ENERGIES OF HETEROCYCLIC COMPOUNDS

A. Introduction

One of the most important applications of thermochemistry is the estimation of the stabilization energies found in molecules with a possibility of π-electron delocalization. This energy of stabilization has been given many different names such as "stabilization energy," "conjugation energy," "delocalization energy," and "resonance energy." Moreover, different authors may have different definitions for these terms. In this context the name "resonance energy" will be used to describe this phenomenon of stabilization, independent of the way it has been estimated. Different methods of estimation sometimes lead to very different results, but it may be difficult to say which of the methods used leads to results that are best related to the true π-delocalization

[100] G. Geiseler and J. Sawistowsky, *Z. Phys. Chem.* (*Leipzig*) **250**, 43 (1972).
[101] G. Geiseler and H.-J. Rauh, *Z. Phys. Chem.* (*Leipzig*) **249**, 376 (1972).
[102] J. D. Cox, *Tetrahedron* **19**, 1175 (1963).

energy. The resonance energies derived from thermochemical data are a composite of many energy contributions, both stabilizing and destabilizing, and hence the true π-delocalization energy may be far from the experimentally determined resonance energy.[103]

It is a common practice to define resonance energy as a difference between the estimated and experimental heats of formation or atomization. The estimated heats of formation or atomization are calculated for the most stable valence-bond structure. Many different bond energy schemes have been in use, with the result that the resonance energies reported by different authors for the same compounds do not agree. Even today, the lack of reliable thermochemical data for many acyclic compounds which are generally regarded as having no resonance stabilization prevents a compilation of a bond energy scheme applicable to many heteroaromatic compounds. It seems unnecessary to increase the confusion around resonance energy values by using any of the available bond energy schemes for a recalculation of resonance energy values for compounds whose heats of formation are given in the literature. Thus the scope of this article is limited to a compilation of the published resonance energy determinations with some discussion of the results.

B. Resonance Energies of Heteroaromatic Nitrogen Compounds

An extensive study of the resonance energies of a number of azoles has been reported by Zimmermann and Geisenfelder.[104] Resonance energy values are given for pyrrole [73], pyrazole [74], imidazole [75], 1,2,4-triazole [76], tetrazole [77], indole [6], indazole [78], benzimidazole [79], benzotriazole [80], and carbazole [81]. The authors exemplified the effect of the chosen bond energy scheme on the calculated resonance

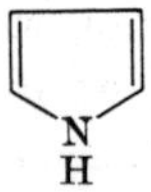

[73] 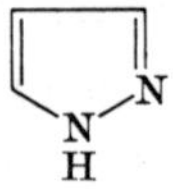   [74]    [75] 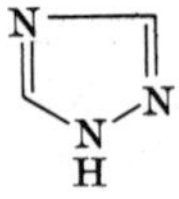   [76] 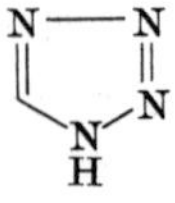   [77]

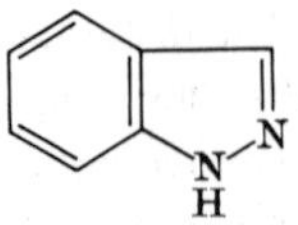

[78] 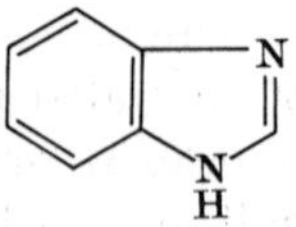   [79] 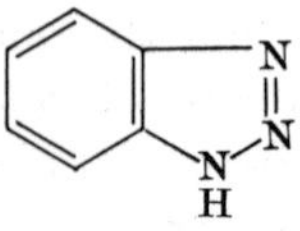   [80] 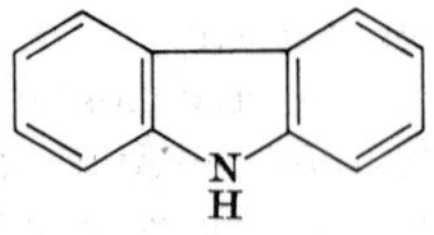   [81]

---

[103] C. T. Mortimer, "Reaction Heats and Bond Strengths," p. 67. Pergamon, Oxford, 1962.

[104] H. Zimmermann and H. Geisenfelder, Z. Electrochem. **65**, 368 (1961).

energy values. Thus, resonance energies ranging from 72.8 to 129 kJ mole^{-1} were obtained for pyrrole using five different bond energy schemes. For *76*, resonance energy values varied between 84 and 206 kJ mole^{-1}. Hence the authors suggested that little attention should be paid to the "absolute" values of resonance energy but that it is much more essential to note the variance of the extent of resonance in a series of homologs using the same bond energy scheme.

The effect of increasing nitrogen atom content in the five-membered azole ring system is revealed by comparing the resonance energy values of *73* [72.8 kJ mole^{-1}), *74* (137 kJ mole^{-1}), *76* (152 kJ mole^{-1}), and *77* (264 kJ mole^{-1}). A continuous increase in resonance energy with increasing nitrogen atom content is observed. The same trend also exists in the benzazoles *6* (175 kJ mole^{-1}), *78* (249 kJ mole^{-1}), and *80* (313 kJ mole^{-1}). According to Zimmermann and Geisenfelder, these facts may be explained as follows. If valence-bond structures are drawn for pyrrole, it is noted that the nitrogen atom always carries a positive charge while partial negative charges occupy α and β positions of the ring. Because nitrogen is more electronegative than carbon, the contributions of the polar valence-bond structures increase and hence the energy content of the molecule decreases as carbon atoms of the pyrrole ring are replaced with nitrogen atoms.

The high resonance energy of *74* relative to *75* is to be attributed to the lesser electrostatic work necessary for charge separation in *74*, where the two nitrogen atoms are in immediate proximity. A study of the resonance energies of *74* and *75* has also been undertaken by Bedford et al.[105] The reported resonance energy values are in good agreement with those given by Zimmermann and Geisenfelder.

Apparently the most recent estimate of the resonance energy of pyridine [*82*] has been reported by Bedford et al.,[98] who give a value of about 134 kJ mole^{-1} for this energy quantity. Using a different method, Tjebbes[106] estimated the resonance energy of *82* to be about 101 kJ mole^{-1}. In this work Tjebbes also estimated resonance energy values

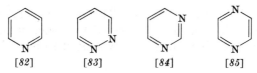

[*82*]   [*83*]   [*84*]   [*85*]

for the three diazines, pyridazine [*83*], pyrimidine [*84*], and pyrazine [*85*]. While *82*, *84*, and *85* have two identical Kekule structures, those

[105] A. F. Bedford, P. B. Edmondson, and C. T. Mortimer, *J. Chem. Soc.* 2927 (1962).
[106] J. Tjebbes, *Acta Chem. Scand.* **16**, 916 (1962).

of *83* are different and hence it was expected that *83* should have a lower resonance energy than *84* and *85*. Contrary to expectations, the resonance energy of *83* (52kJ mole^{-1}) was found to be higher than that of either *84* or *85* (about 34 kJ mole^{-1} in both cases). Surprisingly, on substituting a —N= for a —CH= group, the resonance energy in this ring size decreases by 67 kJ mole^{-1} while the opposite trend was observed by Zimmermann and Geisenfelder[104] in the five-membered heterocycles. These surprising conclusions were criticized by Tjebbes as due to erroneous bond energy values. On the other hand, Bedford *et al.*[98] give a value of about 102 kJ mole^{-1} for the resonance energy of pyrazine [*85*].

The resonance energies of the two pyridine derivatives with fused benzene rings, quinoline [*86*] and acridine [*87*], have been determined

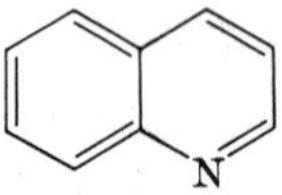

[*86*] 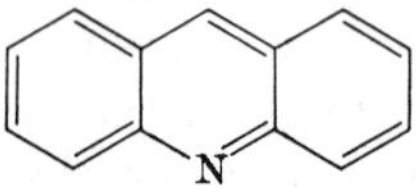   [*87*]

by Klages[107] and Jackman and Packham,[108] respectively. For the former, a resonance energy of 198 kJ mole^{-1}, and for the latter, 350 kJ mole^{-1}, were estimated.

Porphin [*88*], the simplest porphyrin, consists of four pyrrole rings

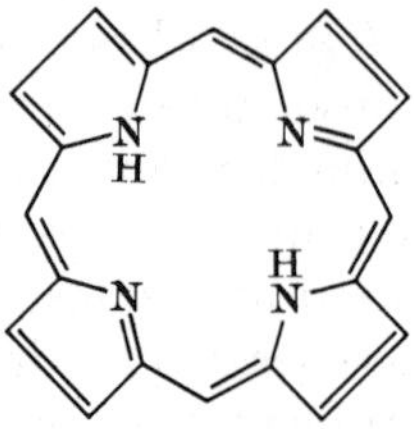

[*88*]

linked together by methine bridges. The heat of combustion of this flat 16-membered ring has been measured by Longo *et al.*[109] The resonance energy of this compound was evaluated to be as high as about $1750 \text{ kJ mole}^{-1}$.

[107] F. Klages, *Chem. Ber.* **82**, 358 (1949).

[108] L. M. Jackman and D. I. Packham, *Proc. Chem. Soc.* 349 (1957).

[109] F. R. Longo, J. D. Finarelli, E. Schmalzbach, and A. D. Adler, *J. Phys. Chem.* **74**, 3296 (1970).

Recently, resonance energies of some nitrogen containing hetero-cycles have been deduced from pK and other equilibrium data.[110a-c]

C. Resonance Energies of Heteroaromatic Oxygen Compounds

The resonance energy of furan [*89*] has been derived[111] both from its heat of hydrogenation to tetrahydrofuran and its heat of combustion.

[*89*]

The difference 72 kJ mole^{-1} between the heats of hydrogenation of furan and 2 moles of cyclopentene was taken to represent the resonance energy of *89*. From the heat of combustion, the value of the resonance energy was found to be 66.1 kJ mole^{-1} (Klages' method) or 93 kJ mole^{-1} (Franklin's method).

Cass *et al.*[112] have determined the resonance energies of dibenzofuran [*90*] and dibenzopyran [*91*] to be about 367 and 382 kJ mole^{-1}, respectively. These values are somewhat higher than the resonance energy

[*90*]   [*91*]

of diphenyl ether (349 kJ mole^{-1}), given by the same authors. The increase in resonance energy relative to diphenyl ether was ascribed to the increase in molecular rigidity in *90* and *91*, assumed to be planar molecules.

D. Resonance Energies of Heteroaromatic Sulfur Compounds

Thiophene [*92*], which is generally regarded as the most aromatic member of the five-membered heterocyclic family furan, pyrrole, and thiophene, has been attributed resonance energies of about 84 and

[110a] D. Lloyd and D. R. Marshall, *Chem. Ind.* 335 (1972).

[110b] M. J. Cook, A. R. Katritzky, P. Linda, and R. D. Tack, *J. Chem. Soc. Perkin Trans. II* 1295 (1972).

[110c] M. J. Cook, A. R. Katritzky, P. Linda, and R. D. Tack, *Tetrahedron Lett.* 5019 (1972).

[111] G. W. Wheland, "Resonance in Organic Chemistry," pp. 86, 99. Wiley, New York, 1955.

[112] R. C. Cass, S. E. Fletcher, C. T. Mortimer, H. D. Springall, and T. R. White, *J. Chem. Soc.* 1406 (1958).

[92]

117 kJ mole^{-1} by Sunner[113] and Wheland,[111] respectively. The relatively high resonance energy of thiophene is probably due to the fact that sulfur is less electronegative than oxygen or nitrogen and is thus capable of releasing electrons into the ring to form a sextet of π electrons required for aromaticity.

Sunner[113] has estimated the resonance energy of thianthrene [93] to exceed the stabilization energy of two benzene rings by 71 kJ mole^{-1}.

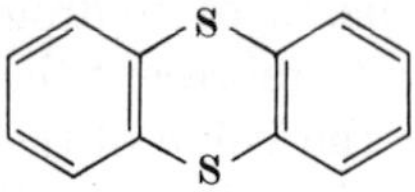

[93] 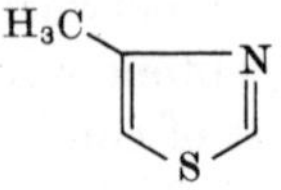   [94] 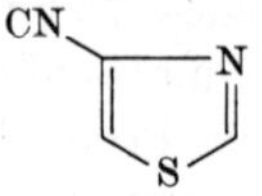   [95]

Månsson and Sunner[114] have determined the heats of combustion of 4-methylthiazole [94] and 4-cyanothiazole [95]. The resonance energy of 94 was estimated to be about 63 kJ mole^{-1}, and that of the unsubstituted thiazole 42–50 kJ mole^{-1}, the same as that of 95.

E. Resonance Energies of Heteroaromatic Phosphorus Compounds

Crude estimates of the resonance energies of pentaphenylphosphole [96] and 9-phenyl-9-phosphafluorene [97] have been reported by

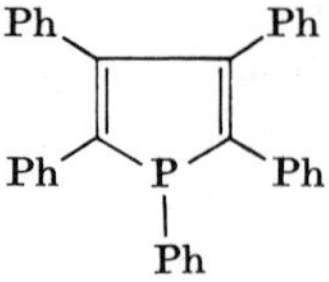

[96] 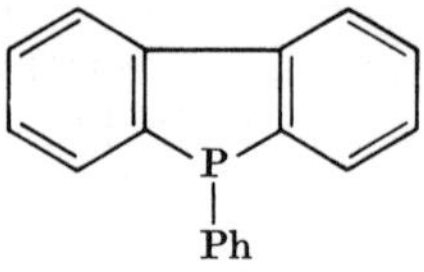   [97]

Bedford et al.[115] The resonance energy of 96 was estimated to be 163 ± 54 kJ mole^{-1} higher than that of its oxide. In addition to the resonance energy of the phenyl groups, 97 was found to have an extra stabilization energy of about 80 kJ mole^{-1}, which probably is to be ascribed to the availability of the $3p$ orbitals of the phosphorus atom, which permits additional electron delocalization.[116]

[113] S. Sunner, Acta Chem. Scand. 9, 847 (1955).

[114] M. Månsson and S. Sunner, Acta Chem. Scand. 20, 845 (1966).

[115] A. F. Bedford, D. M. Heinekey, I. T. Millar, and C. T. Mortimer, J. Chem. Soc. 2932 (1962).

[116] C. T. Mortimer, "Reaction Heats and Bond Strengths," p. 83. Pergamon, Oxford, 1962.

F. Resonance Energies from Thermochemical Data in Relation to Molecular-Orbital Calculations

Recently Dewar et al.[117a-d] and Hess et al.[118] have evaluated resonance energies for a number of aromatic heterocycles using MO calculations. They found that bonds in classic polyenes and classic conjugated compounds containing heteroatoms are localized, in the sense

TABLE XII

RESONANCE ENERGIES OF SOME HETEROCYCLIC COMPOUNDS FROM THERMOCHEMICAL
DATA AND MOLECULAR ORBITAL CALCULATIONS

	Resonance energy (or β value) (kJ mole^{-1})	
Compound	Thermochemical data	MO calculations
Pyrrole [73]	72.8[a]	22.2,[i] 35.7,[j] 0.233β[k]
Imidazole [75]	74.1[a]	64.6[j]
Indole [6]	175[a]	0.466β[k]
Indazole [78]	204[a]	129[j]
Carbazole [81]	331[a]	0.715β[k]
Pyridine [82]	101,[b] 134[c]	96.7,[i] 87.6[j]
Pyridazine [83]	51.5[b]	71.5,[i] 61.3[j]
Pyrimidine [84]	33.5[b]	84.5[j]
Quinoline [86]	198[d]	143,[i] 138[j]
Acridine [87]	351[e]	173[i]
Furan [89]	66.1,[f] 72.0[f]	18.0,[i] 6.7,[j] 0.044β[k]
Dibenzofuran [90]	367[g]	167,[j] 0.655β[k]
Thiophene [92]	84,[h] 117[g]	27.2[l]

[a] H. Zimmermann and H. Geisenfelder, Z. Electrochem. **65**, 368 (1961).

[b] J. Tjebbes, Acta Chem. Scand. **16**, 916 (1962).

[c] A. F. Bedford, A. E. Beezer, and C. T. Mortimer, J. Chem. Soc. 2039 (1963).

[d] F. Klages, Chem. Ber. **82**, 358 (1949).

[e] L. M. Jackman and D. I. Packham, Proc. Chem. Soc. 349 (1957).

[f] G. W. Wheland, "Resonance in Organic Chemistry," pp. 86, 99. Wiley, New York, 1955.

[g] R. C. Cass, S. E. Fletcher, C. T. Mortimer, H. D. Springall, and T. R. White, J. Chem. Soc. 1406 (1958).

[h] S. Sunner, Acta Chem. Scand. **9**, 847 (1955).

[i] M. J. S. Dewar and N. Trinajstić, Theor. Chim. Acta (Berlin) **17**, 235 (1970).

[j] M. J. S. Dewar, A. J. Harget, and N. Trinajstić, J. Amer. Chem. Soc. **91**, 6321 (1969).

[k] B. A. Hess, Jr., L. J. Schaad, and C. W. Holyoke, Jr., Tetrahedron **28**, 3657 (1972).

[l] M. J. S. Dewar and N. Trinajstić, J. Amer. Chem. Soc. **92**, 1453 (1970).

[117a] M. J. S. Dewar and N. Trinajstić, Theor. Chim. Acta (Berlin) **17**, 235 (1970).

[117b] M. J. S. Dewar, A. J. Harget, and N. Trinajstić, J. Amer. Chem. Soc. **91**, 6321 (1969).

[117c] M. J. S. Dewar and G. J. Gleicher, J. Chem. Phys. **44**, 759 (1966).

[117d] M. J. S. Dewar and N. Trinajstić, J. Amer. Chem. Soc. **92**, 1453 (1970).

[118] B. A. Hess, Jr., L. J. Schaad, and C. W. Holyoke, Jr., Tetrahedron **28**, 3657 (1972).

that the calculated heats of atomization can be expressed as sums of "polyene" bond energies. Hence the above authors define resonance energy as the difference between the heat of atomization of a given compound and the heat of atomization calculated for a corresponding classic structure using the "polyene" bond energies. Table XII shows a collection of resonance energies for a few heteroaromatic compounds calculated according to these principles. For comparison, resonance energies derived from thermochemical data using the classic procedure have also been included in the table. It is seen that resonance energies based on polyene bond energies are generally much smaller than those calculated using the classic method. Thus, the "Dewar" resonance energies of furan, pyrrole, and thiophene are rather low compared with the resonance energy of benzene (92 kJ mole^{-1}), which shows that the aromatic character of these heterocycles may not be as high as is generally believed.

Author Index

Numbers in parentheses are reference numbers that are included for ease of identifying a reference when the author's name is not cited in the text or is not the first-named author in a full reference citation.

A

Aaljoki, K., 214(47), 215(47)
Aarons, L. J., 126(143)
Abate, K., 168, 176(34), 179(34), 183(34, 53)
Adam, F. C., 122(127), 123(127), 126(143)
Adam, W., 14(54, 55)
Adams, R. F., 117(75)
Adler, A. D., 242(109)
Adler, T., 169
Äyräs, P., 217(55, 58a)
Aguiar, A. M., 137(194)
Akahori, Y., 122(123)
Akulinin, O. B., 70(h), 71, 75(o), 77
Al-Baldawi, S. A., 120(96, 97)
Albert, A., 194(73)
Alberti, A., 144(247)
Albridge, R. G., 1(20)
Alderweireldt, F., 210(34a, 34b)
Aleksandrov, A. N., 63(29)
Alger, R. S., 96, 98(4)
Al-Joboury, M. I., 2(21)
Allen, T. L., 204, 206(c), 209(13)
Allendoerfer, R. D., 140(213), 141(213)
Allinger, N. L., 209(30)
Almenningen, A., 64(46)
Alt, H., 13(50), 36(50)
Altona, C., 221(74a)
Ambros, D., 224(79a), 225, 226(79a)
Ames, D. L., 10(36), 17(36), 39(136)
Ames, D. P., 63(39, 40), 91(161)
Andersen, J. R., 70(d, f, g, j, k), 71(o), 73(60), 75(l, s, t, u), 76(oo, pp), 77(67), 78, 87(135)

Anderson, N. H., 108(37)
Andresen, U., 79(81)
Angyal, S. J., 209(30)
Anteunis, M., 210(34a, 34b, 35a, 35b), 311(d), 217(60, 61, 62), 219(66, 71b, 71c, 71d)
Anteunis-De Ketelaere, F., 217(60)
Armarego, W. L. F., 121(108), 148(4)
Arnold, W., 77(66), 80(66)
Arthur, J. C., 107(31)
Åsbrink, L., 15(61, 62, 63, 65, 66), 34(119, 122), 39(62, 63), 41(65), 42(66)
Ashe, A. J., III, 11(44), 36(44), 37(44), 87, 88(138)
Asmussen, E., 71(o), 73(60), 76(oo), 78
Atherton, N. M., 117(75, 78), 118(84), 119(92, 93, 94), 120(92, 93, 94), 122(122)
Atlani, P., 145(267)
Atwell, W. H., 129(161, 163)
Ayscough, P. B., 96, 98(2)

B

Badoche, M., 202(k), 203, 207, 232(g)
Baer, Y., 2(24)
Bahurel, Y., 221(73a)
Bain, A. D., 32(113), 33(113, 118, 165), 39(165), 48(118)
Bak, B., 60, 70(d, g, k), 71, 75(h, t, u), 76(pp), 77, 78, 79(78), 87(135)
Baker, A. D., 1(1, 10, 17), 3(17), 11(40), 23(1), 24, 32(110), 33(115, 116), 34(40,

247

Cumulative Author Index, Volumes I–VI

Boldface roman numerals indicate volume numbers.

Cumulative Subject Index,
Volumes I–VI

Boldface roman numerals indicate volume numbers.

A

Acetylacetonates
GLC, **III**, 395
Acridines
dipole moments, **IV**, 256
ESR spectra, **VI**, 118
IR spectra, **II**, 300, 323, 324; **IV**, 402, 403, 433
mass spectra, **III**, 283
phosphorescence, **VI**, 165
pK_a, **I**, 20, 65; **III**, 8
reduction, **I**, 270
resonance energy, **VI**, 242, 245
solubility, **I**, 179
UV spectra, **II**, 80; **III**, 161–163; **VI**, 185
X-ray structures, **V**, 425, 453
Acridone
pK_a, **III**, 8
Acylamides
tautomerism, **I**, 41
Adenines
dipole moments, **IV**, 249
IR spectra, **II**, 302
X-ray structures, **I**, 171, 172; **V**, 127, 139, 140, 145, 156, 157, 297, 299, 300, 301, 306, 314, 315
Adenosines
dipole moments, **IV**, 249
ESR spectra, **VI**, 112
MCD spectra, **III**, 422
X-ray structures, **V**, 72, 77, 80, 94, 100, 133, 151, 162, 299, 302, 303, 305, 311, 519
Adenylic acids
crystal structure of, **I**, 171–172

Adrenochrome
half-wave potential, **I**, 290
Alizarine blue
oxidation-reduction potential, **I**, 224
Alkaloids, *see also* individual alkaloids
catalytic currents, **I**, 233
determination of hydrolysis, **I**, 278
GLC, **III**, 346–352
retention times, **III**, 347, 348
half-wave potential, **I**, 312–314
IR spectra, **II**, 213
NMR spectra, **II**, 153
polarography of , **I**, 272
Alloxans
crystal structure of, **I**, 171–172
ESR spectra, **VI**, 127, 142
oxidation-reduction potential, **I**, 285
potentiometry, **I**, 249
X-ray structures, **V**, 128, 135
Alloxazines
X-ray structures, **V**, 430–434, 485
Angiotensin
ESR spectra, **VI**, 113
Anhydrides, cyclic
IR spectra, **II**, 188
Annulenes
ESR spectra, **VI**, 118
IR spectra, **IV**, 304
NMR spectra, **IV**, 171
Annulenetrisulfide
ESR spectra, **VI**, 124
Anthocyanins
half-wave potential, **I**, 305
Anthracenes
fluorescence, **VI**, 156, 159
phosphorescence, **VI**, 156, 165

297

301